Current Topics in Microbiology and Immunology

Ergebnisse der Mikrobiologie und Immunitätsforschung

58

Edited by

W. Arber, Basle · W. Braun, New Brunswick · R. Haas, Freiburg · W. Henle, Philadelphia · P. H. Hofschneider, Munich · N. K. Jerne, Basle · P. Koldovský, Philadelphia · H. Koprowski, Philadelphia · O. Maaløe, Copenhagen · R. Rott, Gießen · H. G. Schweiger, Wilhelmshaven · M. Sela, Rehovot · L. Syruček, Prague · P. K. Vogt, Los Angeles · E. Wecker, Würzburg

With 9 Figures

Springer-Verlag Berlin · Heidelberg · New York 1972

ISBN-13: 978-3-642-65359-9 e-ISBN-13: 978-3-642-65357-5
DOI: 10.1007/978-3-642-65357-5

Softcover reprint of the hardcover 1st edition 1972
 Universitätsdruckerei H. Stürtz AG, Würzburg

Table of Contents

List of Contributors

Professor E. Benjamini, Ph. D., Department of Medical Microbiology, University of California, School of Medicine, Davis, CA 95616/USA

Dr. Robert C. Boguslaski, Ames Research Laboratory, Miles Laboratories Inc., Elkhart, IN 46514/USA

Professor Myron Levine, Ph. D., The University of Michigan, School of Medicine, Department of Human Genetics, Ann Arbor, MI 48204/USA

Dr. Mirko Majer, Swiss Serum and Vaccine Institute Berne, P. O. Box 2707 CH-3001 Berne

Dr. N. S. Mhatre, Managing Director, Miles-Yeda Ltd. Kiryat Weizmann, P. O. Box 1122, Rehovoth, Israel

Dr. D. Michaeli, Assistant Professor, Department of Biochemistry and Surgery, University of California, School of Medicine, San Francisco, CA 94122/USA

Bruce A. Phillips, Assistant Professor of Microbiology, School of Medicine, University of Pittsburgh, Pittsburgh, PA 15213/USA

Dr. Richard S. Smith, Miles Laboratories, Inc., 1127 Myrtle Street, Elkhart IN 46514/USA

Dr. J. D. Young, Adjunct Associate Professor, Department of Bacteriology and Immunology, University of California, Berkeley, CA 94720/USA

Applications of Bound Biopolymers in Enzymology and Immunology

ROBERT C. BOGUSLASKI[1], RICHARD S. SMITH[2], and NAGESH S. MHATRE[3]

Contents

I. Introduction

Bound biopolymers have been extensively investigated, especially in the last ten years, and other reviews of some of this work have appeared (I. SILMAN and KATCHALSKI, 1966; WELIKY and WEETALL, 1965; GUILBAULT, 1968; BARKER and EPTON, 1970; MOSBACH, 1971). Both soluble and insoluble complexes of bound biopolymers have been studied (I. SILMAN and KATCHALSKI, 1966; NAKANE and PIERCE, 1966). The interest in these materials stems from the

1 Ames Research Laboratory and 2 Science Information Services, Miles Laboratories, Inc., Elkhart, Indiana, 46514, U.S.A.
3 Miles-Yeda Limited, Rehovot, Israel.

unusual properties they exhibit (L. GOLDSTEIN et al., 1964; GOLDMAN et al., 1968b), the fact that they may serve as models for intracellular enzyme systems (I. SILMAN, 1969), and their exceptional potential for industrial application (SUZUKI et al., 1967; TSUMURA and ISHIKAWA, 1967; WILSON and LILLY, 1969; BARKER and EPTON, 1970).

The term bound biopolymer, as used here, refers to an enzyme, antigen, antibody, or polynucleotide which is linked, usually through covalent bonds, to a carrier. This carrier may be another biopolymer (NAKANE and PIERCE, 1966) or it may be a synthetic matrix material (KAY, 1968). The carrier may be soluble or insoluble; thus the complex produced may be soluble or insoluble.

Proteins or polynucleotides have been coupled to carriers by a variety of methods. These ordinarily fall into three categories (I. SILMAN and KATCHALSKI, 1966). The polymers are bound to the carriers by adsorption (McLAREN and PACKER, 1970; GOLDMAN et al., 1971a), through covalent bonds (I. SILMAN and KATCHALSKI, 1966), or by entrapment within a polymer matrix (BERNFELD and WAN, 1963; GUILBAULT, 1968). The advantages and disadvantages of each method have been discussed in the literature (I. SILMAN and KATCHALSKI, 1966; TOSA et al., 1966a; BERNFELD and WAN, 1963) and will not be reiterated in this review.

There have been a number of reviews in recent years in the field of insoluble enzymes, antigens and antibodies. MANECKE (1964, 1968) reviewed work on insoluble enzymes. I. SILMAN and KATCHALSKI (1966) discussed the preparation and properties and some of the applications of water insoluble enzymes and antigens. GUILBAULT (1966, 1968) included short sections on insoluble enzymes in his reviews on the analytical uses of enzymes. The preparation and properties of cellulose derivatives, and their use in engendering insoluble immunochemical or enzyme systems were reviewed by WELIKY and WEETALL (1965) and by CROOK et al. (1970). CHIBATA and TOSA (1966, 1967) and TOSA and CHIBATA (1967) presented three reviews on the preparation, properties and uses of insoluble enzymes. CROOK (1968) briefly discussed five methods for preparing insoluble enzymes, and KAY (1968) recently prepared a short review on this topic. A recent article by L. GOLDSTEIN and KATCHALSKI (1968) dealt mainly with the effects of the carrier on the kinetic characteristics of immobilized enzymes, as does a recent review by L. GOLDSTEIN (1970). These reviews were generally oriented toward the preparation and properties of insoluble biopolymers, as well as the effects which the carriers have on the polymer's properties. Although many had short sections on the uses of these materials, this area is so large and has such great potential that the applications of these materials deserve further and more complete exposure.

Bound biopolymers have been adapted in a number of basic studies and have shown industrial utility. These derivatives have been proposed as models of intracellular enzymes (GOLDMAN et al., 1968b; McLAREN and PACKER, 1970), and as models of metabolic cycles (BROWN et al., 1968; MOSBACH and MATTIASSON, 1970). They have also been used to isolate antigens and antibodies (I. SILMAN and KATCHALSKI, 1968; PORATH, 1968), in the study of protein structure (L. GOLDSTEIN and KATCHALSKI, 1968), for histochemical localization of antigens (NAKANE and PIERCE, 1966; AVRAMEAS, 1970), and even therapeutically (CHANG and POZNANSKY, 1968; CHANG, 1971).

This review will describe general applications of bound biopolymers. There will be examples of the uses of these materials in basic and applied areas. Separate sections of this review will be devoted to the manner in which bound enzymes, antigens and antibodies have been utilized. It will also mention uses of polymerized antigens and antibodies, and soluble complexes of coupled biopolymers. Short sections on affinity chromatography and the uses of bound biopolymers in nucleic acid research have been included since they offer some very interesting directions for future research. There will be no attempt to cover all the ways in which bound biopolymers have been exploited; however, a number of examples of the application of these materials in a variety of situations will be discussed.

II. Bound Enzymes

A. Naturally Bound Enzymes

Most biologically active proteins, *in vivo*, are bound in some way to cellular material. The surroundings provided by the structural material of a cell may influence the properties of some of the embedded protein; thus, the properties exhibited by the protein in solution may not necessarily reflect the properties of these proteins in their natural environment (I. SILMAN, 1969). In addition, D. E. GREEN and JARNEFELT (1959) stated that the sum of the individual activities of the enzymes of a composite system may not always be equivalent to the enzyme activity of the integrated system. This phenomenon could be due, at least in part, to the role structure plays in the functioning of an enzyme system. Thus, a great deal of interest has been generated in determining the relationship between structure and the activity displayed by bound biopolymers (MCLAREN and PACKER, 1970; GOLDMAN et al., 1971a). In addition, KATCHALSKI (1970) recently discussed a synthetic approach for studying microenvironmental effects on enzyme action.

The effect of structure on the properties of proteins may be determined by studying the protein while it is still attached to its natural surroundings, i.e. by studying particular matter. An alternate approach is to construct and investigate models of bound enzyme systems.

Many investigators have examined the properties of naturally insoluble biopolymers. These naturally insoluble systems should display properties which differ from their characteristics in solution if binding does affect the traits of the biopolymer. In some instances, there were distinctive differences in properties between the soluble and the insoluble materials. Only a few of the many examples in the literature will be given here to illustrate these points.

AKATSUKA and NELSON (1966) found the behavior of starch granule-bound ADP glucose-starch glucosyltransferase isolated from embryos of maize seeds was considerably different from that of a similar preparation from endosperms. The authors attributed these differences to different enzyme systems, although they caution that the structure or microenvironment that the granules provide for the enzyme must not be disregarded. Further studies on this enzyme (FRYDMAN and CARDINI, 1967) indicated that granule structure may play an important part in determining the characteristics of the bound enzyme. It was found that certain properties of the enzyme were altered when the particulate enzyme was solubilized. In fact, the specificity of the bound enzyme for UDP-glucose was completely lost on solubilization.

RAO et al. (1968) demonstrated hyperbolic substrate-velocity curves for the alkali cation-activated AMP deaminase bound to the erythrocyte membrane, whereas the soluble enzyme produced a sigmoidal curve.

The pH optima for soluble and mitrochondria-bound aconitase and glutamic dehydrogenase of *Lupinus albus* were investigated (ESTERMANN et al., 1959). The bound aconitase demonstrated a shift in the pH optimum of 1.2 units, whereas glutamic dehydrogenase did not demonstrate a change. Membrane-bound acetylcholinesterase in a subcellular fraction from *Electrophorus electricus* exhibited a pH dependence, in the absence of buffer, which differed from that of the soluble enzyme (H. I. SILMAN and KARLIN, 1967). This discrepancy was much less apparent when the enzyme was assayed in the presence of buffer, and solubilization of the membrane-bound enzyme eliminated any anomalous pH behavior in the presence or absence of buffer. The authors ascribe these effects to a difference between the local pH in the membrane, due to substrate hydrolysis, and the pH of the bulk solution. A similar explanation was previously proposed by MCLAREN (1957) for the anomalous pH dependence of chymotrypsin adsorbed on the surface of kaolinite particles as opposed to the same enzyme in solution. KATZ and MAYER (1969) found significant differences in the K_m and V_{max}, response to sodium chloride, and pH optima between catechol oxidase in sugar beet chloroplasts and the solubilized enzyme.

In contrast to the preceding, other investigators (ARNOLD, 1966; YUASA et al., 1967) found that the bound and soluble forms of the enzymes they were investigating behaved similarly.

B. Enzymes Acting at Surfaces

Evidently, the environment of a bound biopolymer and, perhaps, the conformation of the bound protein may alter the properties of this material. Identification of the features responsible for these alterations would have wide implications in understanding the functioning of proteins in subcellular systems.

A large number of investigations were carried out in which enzymes, as representative biopolymers, were fixed in or to materials of defined structure and the properties of the bound enzymes were determined. MCLAREN and PACKER (1970) have reviewed the extensive work of MCLAREN and co-workers on the action of enzymes in structurally restricted systems. They concluded that, in many cases, structurally restricted enzymes possess fundamental properties which differ from the properties of enzymes in solution. Certain structures produce environments for the bound protein in which the hydrogen ion concentration is different from that in bulk solution. Thus, structure can be responsible for modifying the properties of an enzyme (McLaren, 1969; McLaren and Babcock, 1959). An interesting aspect of their work involved a study of the hydrolysis of insoluble substrates by soluble enzymes (McLaren and Packer, 1970). These systems may serve as models for digestive processes.

Vorob'Eva and Poltorak (1966b) demonstrated that the adsorption of hexokinase to certain hydrophobic surfaces had little effect on its activity, whereas a 16–40 fold decrease in activity was noted when this enzyme was adsorbed to hydrophilic surfaces. Likewise, in studies of model systems for biological membranes, differences in enzyme activity were encountered when the enzyme was placed in various environments. Catalase decreased in activity on adsorption to hydrophilic surfaces, but increased in activity when adsorbed on hydrophobic surfaces (Gol'dfel'd et al., 1966). The activity of alkaline phosphatase in monolayers of lauric acid or cephalin adsorbed on either activated carbon or silica gel increased about six times over that of the solution value (Poltorak and Vorob'Eva, 1966). The activity of acid phosphatase and phosphoglucomutase decreased on adsorption to silicon dioxide, but the K_m values of these enzymes remained unchanged (Vorob'Eva and Poltorak, 1966a). In the same paper, acid phosphatase adsorbed on silicon dioxide coated with cephalin was shown to retain all its activity, whereas the same enzyme on carbon coated with cephalin demonstrated a 20% decline in activity. In addition, phosphoglucomutase adsorbed on silicon dioxide coated with cephalin lost all activity, but the enzyme on carbon coated with cephalin lost only one-half its activity. Alkaline phosphatase adsorbed to a silicon dioxide-cephalin membrane was 24 times more active than in solution, and this increase apparently was due to an increase in the rate constant for the breakdown of the enzyme substrate complex (Poltorak and Vorob'Eva, 1967).

It is apparent that structure may influence the properties of biopolymers. Thus, by placing biopolymers in structures that provide known environments, some knowledge as to the individual effects produced by a particular matrix may be provided. Recently, a number of synthetic membranes were prepared with enzymatic activity (Broun et al., 1969; Goldman et al., 1968b, 1971). The study of the properties of these model membranes may aid in the understanding of the interrelationships of structure and enzyme properties in biological membranes, and the kinetic nature of membrane bound enzymes.

C. Application of Bound Enzymes

1. Membrane Models

A papain collodion membrane was prepared and its properties investigated (Goldman et al., 1965, 1968a, 1968b, and to be published; I. Silman, 1969). This enzyme membrane was prepared by absorbing papain into a collodion matrix and then crosslinking the absorbed enzyme. In this manner, the thickness of the enzyme layer in the membrane could be controlled and a variety of membranes could be prepared. Thus, membranes consisting of one enzyme layer, one enzyme layer and an adjacent collodion layer, and two papain layers separated by a collodion layer were available. A quantitative analysis of the absorption process was also presented. In addition, such physical parameters as the filtration and permeability coefficient were determined, and were used to calculate the pore radii (278–308 A°) of the membrane (GOLDMAN et al., 1968b).

Three layer membranes containing non-activated papain retained 90% of the enzyme activity when stored for 18 months in 0.05 M, pH 7.6 phosphate buffer at 4°C. Membranes containing activated papain and stored in a similar manner in the presence of cysteine and EDTA lost more than half of their activity in the same time period. The activated three layer membrane was somewhat less thermally stable at high temperature (65–80°C) than the solution enzyme. However, the papain-collodion membrane was more stable than the native enzyme in the low pH region (1–2.75) and retained full activity, as does the native enzyme, at high pH values (GOLDMAN et al., 1968b).

The pH-activity profiles of the three layer membrane were determined for a variety of substrates, including benzoyl-L-arginine ethyl ester, benzoyl-glycine ethyl ester, benzoyl-DL-arginine p-nitroanilide, benzoyl-L-arginine amide and acetyl-L-glutamic acid diamide (GOLDMAN et al., 1968b; I. SILMAN, 1969). The pH-activity curves determined with these substrates differed from one another and from the bell-shaped curve for the solution enzyme. Optimum activity was always displaced toward higher pH values for the enzyme membranes, the differences being most readily apparent for good substrates and much less apparent for poor substrates. The authors suggested that these discrepancies were caused by differences between the local pH in the membrane and the pH of the bulk solution. As the enzymic reaction proceeds, protons are produced and accumulate in the membrane, making the pH in the membrane a few units lower than in the external solution. Thus, the pH optimum for the enzyme appears to occur at a higher value. Of course, the greater the rate of reaction, the more marked the apparent difference in pH optima.

There have been three main lines of proof offered to support the reasoning that local pH changes are responsible for the anomalous pH-activity behavior of the membrane. A pH indicator, neutral red, added to a solution containing substrate and an inactive non-crosslinked enzyme membrane produced both a yellow membrane and yellow solution at pH values above 7.0. Upon activation of the papain, the membrane turned red at external pH values up to 10.0, while the solution remained yellow. Also, buffers of high ionic strength tend to abolish any discrepancy in the pH-activity curves. Finally, papain-collodion membranes frozen and powdered show pH-activity curves more like the soluble enzyme (GOLDMAN et al., 1968b; I. SILMAN, 1969).

Similar pH differences were also found in the comparison of soluble and naturally bound enzymes (ESTERMANN et al., 1959; H. I. SILMAN and KARLIN, 1967; I. SILMAN, 1969; KARLIN, 1965). For example KARLIN (1965) found marked differences in the pH activity curves between soluble and membrane bound acetylcholinesterase from the electric organ of *Electrophorus electricus*.

A theoretical analysis of the kinetics of an enzyme embedded in a membrane has also been presented (GOLDMAN et al., 1968a; GOLDMAN et al., 1971a), assuming the membrane is at steady state and the substrate and product concentrations in the membrane are determined by diffusion and reaction rates. Equations were derived for the sum of flows of substrate and product, and the sum of substrate and product concentrations anywhere in the membrane under a variety of boundary conditions. Expressions for the separate flows and concentration gradients of substrate and product in the membrane were developed, assuming first order kinetics for the local enzymic reaction. In addition, an expression for the overall reaction rate of an enzyme membrane was derived.

The above analysis was performed disregarding any effect of the Nernst diffusion layer. More recently, GOLDMAN et al. (1971) described the kinetic behavior of alkaline phosphatase-collodion membranes and gave a theoretical analysis which showed the effect of this layer on the apparent kinetic behavior of the enzyme membrane system. This diffusion layer, along with the catalytic parameters of the system, help control the concentration of substrate at the membrane-solution interface and, in turn, the overall rate of reaction. SUNDARAM et al. (1970) presented a theoretical analysis of an enzyme membrane in contact with substrate solution at each interface, and examined the effect of partitioning and diffusion on this system. Some computer solutions were given to illustrate certain principles.

Investigations into the action of enzymes in membranes led SELEGNY et al. (1968) into impregnating chromatographic paper or cellophane with an enzyme, then crosslinking it with bis-diazotized-o-dianisidine. The kinetic behavior of a glucose oxidase membrane was investigated and used to verify an equation, based on diffusion and reaction, which described the system. Glucose oxidase bound to cellophane membranes (BROUN et al., 1969; SELEGNY et al., 1969a) had the same pH optimum, Michaelis constant, and specificity as the soluble enzyme. However, the matrix bound material appeared to possess greater thermal stability (BROUN et al., 1969). Carbonic anhydrase has been covalently linked to a silastic membrane and used as a model system for the facilitated transport of carbon dioxide across membranes (BROUN et al., 1970).

There have been other studies focusing on the kinetic behavior of bound enzymes (L. GOLDSTEIN et al., 1964; HORNBY et al., 1966, 1968; LILLY et al., 1966; WILSON et al., 1968a, 1968b; L. GOLDSTEIN and KATCHALSKI, 1968; WHARTON et al., 1968; SELEGNY et al., 1969b). The kinetic behavior of glucose oxidase bound to porous glass particles has been studied by WEIBEL and BRIGHT (1971). Also, there have been investigations into the kinetic properties of crystalline enzymes. DOSCHER and RICHARDS (1963) discussed the problems of diffusion limitation in the determination of kinetic parameters for crystalline enzymes. They investigated the kinetic behavior of crystalline ribonuclease S and found it to be very similar, but not identical to the kinetics displayed by an amorphous suspension of the enzyme. However, both of these values differed from solution values. SLUYTERMAN and DE GRAFF (1969) examined the activity of two crystal modifications of papain and found that substrate breakdown by the crystals was governed by both diffusion and reaction rate. They effectively removed diffusion as a consideration and found that activity of both crystalline forms was equal to the solution enzyme.

In addition to the effects of the matrix, the properties of the enzyme itself may be altered by a conformational change when it is bound. BROWN et al. (1966b) suggested that the mode of action of ATPases from either the endoplasmic reticulum of barley root or the sarcoplasmic reticulum of rabbit heart can be derived from the type of binding which the protein undergoes when bound to the native membrane or synthetic support. They further indicated (BROWN et al., 1966a) that properties of soluble and membrane-bound rabbit heart sarcoplasmic reticulum ATPase should differ if the membrane's structure plays a role in determining the enzyme's characteristics. If the differences in the properties of the soluble and bound enzyme are caused by a change in enzyme conformation, then coupling the soluble enzyme to a synthetic matrix may restore some of the properties it possessed while bound in the biological membrane. The membrane ATPase was solubilized and its properties were compared to those displayed by the membrane fraction, and to those of the enzyme bound to a synthetic matrix. The soluble enzyme had a pH-activity optimum at 7.2, displayed only a small and variable response to ouabain, and showed no increase in activity upon the addition of sodium or potassium. The membrane-bound fraction displayed a pH-activity optimum at 8.1, was stimulated by sodium and potassium, and was inhibited by ouabain. The matrix-bound enzyme had a pH-activity optimum at 7.2, was inhibited by ouabain, and showed no response to sodium or potassium at the levels used with the membrane fraction (BROWN et al., 1966a). Thus, the properties of the naturally bound ATPase may be due to a three dimensional structure which is lost on solubilization. The native conformation of the enzyme may be partially restored on binding to the synthetic support, since this material exhibited some of the same properties as the membrane bound material.

A similar study was performed with an ATPase from the membrane fraction of barley root (BROWN et al., 1967). Again, there were discrepancies in the properties of the membrane-enzyme, the matrix-bound material, and the solubilized protein. The membrane-bound enzyme showed the greatest sensitivity to ouabain, followed by the soluble enzyme, and then the supported enzyme. Potassium greatly stimulated the matrix-bound preparation, but only slightly affected the naturally bound material. A functional interrelationship between the enzyme and the membrane was again demonstrated.

Other investigators (WHITTAM et al., 1968; WHEELER et al., 1969) also prepared and investigated membranes possessing bound ATPase. Although they were successful in preparing active derivatives of potato apyrase bound to cellulose, filter paper, dialysis tubing, collodion membranes, and Millipore filters, these same methods were unsuccessful with $(Na^+ + K^+)$—ATPase. The cellulose-bound apyrase derivatives had properties similar to the soluble enzyme, except that the bound enzymes had a greater K_m value and a lower V_{max} value than the solution enzyme (WHEELER et al., 1969).

BLUMENTHAL et al. (1967) proposed a model system for active transport by constraining a solution of papain between two oppositely charged ion exchange membranes and using an uncharged, low molecular weight substrate. Enzymatic hydrolysis of the substrate produced ammonium ions, and a potential difference arose when electrodes reversible to this ion were placed in the system. Thus, coupling of an enzymatic reaction to electric current flow was achieved.

SHARP et al. (1969) employed cellulose and DEAE-cellulose sheets as support matrices for β-galactosidase. The kinetic behavior of these derivatives was described by an equation similar to the Michaelis-Menten equation. However, the kinetic parameters were dependent on the thickness of the liquid diffusion layer over the sheets. The data on the long term use of these sheets was encouraging, and they may have application in chemical reactors. Similarly, pyruvate kinase (WILSON et al., 1968b), creatine kinase, and lactate dehydrogenase (KAY et al., 1968) have been bound to cellulose and studied.

2. Cytochemical Models

An interesting application of bound enzymes occurred in the field of cytochemistry. MITZ (1956) examined catalase bound to the surface of an immobile phase (anion exchange cellulose) adjacent to a mobile phase (substrate in solution) as a model for the interior of a cell. Similarly, alkaline phosphatase was entrapped in a polyacrylamide film (VAN DUIJN et al., 1967) and this system was used as a model for investigating cytochemical reactions of this enzyme. The influence of substrate diffusion on the kinetic behavior of the enzyme in a cytochemical system was treated theoretically. The influence of several parameters on the diffusion of substrate, and the conditions whereby the amount of dye precipitated would be proportional to enzyme activity in the film were investigated. These studies indicated the possibilities of cytochemical quantification of enzymes in biological systems. An attempt at quantification of alkaline phosphatase in the leukocytes of a guinea pig exudate, using polyacrylamide films containing a sonicate of the leukocytes as a reference, has recently been attempted (VAN DER PLOEG and VAN DUIJN, 1968). A relationship between the biochemical and cytochemical activities of the enzyme was determined for the films. This proportionality was then used to calculate the quantity of alkaline phosphatase present in the cell, based on its cytochemical activity. The cytochemical staining method detected only 60% of the enzyme activity when compared to sonicates of the leukocytes. The remainder may be in granules not readily permeated by the cytochemical reagents.

There have been a number of miscellaneous uses of enzyme films. SCHWABE (1969) used aminopeptidase on a supported calcium phosphate gel as a membrane for hydrolysis of various peptides. He suggests the use of such enzyme films in the determination of amino acid composition and sequencing of small peptides. MAZIA and HAYASHI (1952) used films of ovalbumin and pepsin in an effort to investigate the effect of structure on a biochemical reaction. OPARIN et al. (1957) found the activity of trypsin appreciably magnified when it was complexed with ergosterol, reinforcing the idea that enzyme activity in a cell may be regulated by interaction with other cell components. ROMEO and DE BERNARD (1966) found that a membrane composed of a structural protein, a phospholipid mixture and lysozyme, which compositionally resembles known lipoprotein membranes, masked a considerable portion of the enzyme's activity.

A number of Russian workers have been doing interesting work in an area related to bound enzymes. They have extensively studied enzyme activity in coacervate droplets as possible cell models (SEREBROVSKAYA and VASIL'EVA, 1964, 1966; OPARIN and SEREBROVSKAYA, 1958, 1966; OPARIN, 1957; SEREBROVSKAYA, 1964; SEREBROVSKAYA et al., 1968; OPARIN et al., 1968; EVREINOVA and BAILEY, 1968). CHANG (1964, 1967) has prepared semipermeable microcapsules which can also serve as cell models.

3. Models of Metabolic Cycles

Hexokinase, phosphoglucoisomerase, phosphofructokinase, and aldolase of the glycolytic sequence were separately entrapped in polyacrylamide gel (BROWN et al., 1968). These gels were packed into a column in the same sequence as the reactions occur in the metabolic cycle. The column was charged with a solution containing glucose, ATP and $MgCl_2$, and the product of the reaction sequence, glyceraldehyde-3-phosphate, was recovered in the eluate.

Pyruvate kinase coupled to cellulose sheets was used in conjunction with lactate dehydrogenase, similarly insolubilized, to carry out a coupled enzymatic reaction; phosphoenolpyruvate was converted to pyruvate, which in turn, was converted to lactate (WILSON et al., 1968b). Recently MOSBACH and MATTIASSON (1970) covalently bound hexokinase and glucose-6-phosphate dehydrogenase to the same polymer matrix so that the product of the first reaction served as the substrate for the second enzyme. In the initial stage of the reaction, the matrix bound enzymes were twice as effective a catalytic system as a solution containing the two enzymes.

4. Protein Structure Studies

a) Immunoglobulins

CEBRA et al. (1961) found they could use a water-insoluble papain derivative to hydrolyze rabbit γ-globulin, without the addition of cysteine to the mixture. The cysteine-free, water-insoluble papain was able to digest γ-globulin without reduction of the globulin or products. Immunoglobulin hydrolyzed in this way retained its sedimentation coefficient and its ability to precipitate antigen and fragmented only after it was reduced. Thus the use of bound papain helped show that the mechanism of digestion of rabbit γ-globulin by cysteine-activated papain, to fragments like those described by PORTER (1959), consists of cleavage succeeded by reduction. Antiovalbumin treated with bound papain gave similar results (CEBRA et al., 1961; CEBRA, 1962). Antigen-antibody precipitates containing antibody from the bound papain treatment dissolved on reduction, giving soluble fragments composed of antigen and portions of antibody (CEBRA et al., 1961).

Rabbit γ-globulin, briefly treated with the insoluble papain and then reduced, gave three fragments. Determination of terminal amino acids in these fragments showed a greater amount of glycine than was found by hydrazinolysis of the globulin (H. I. SILMAN et al., 1962). Rabbit γ-globulin and antibodies to lysozyme and poly-L-tyrosyl gelatin were subjected to limited treatment with bound papain, and then reduced (GIVOL and SELA, 1964). Additional treatment, including digestion and peptide mapping, produced peptide maps for fragments I and II which were different than the map for fragment III, and showed small but distinct differences from each other. The peptide maps of fragment III from either rabbit γ-globulin or antibody were essentially the same.

Immune rabbit γ-globulin briefly treated with bound papain then with sodium dodecylsulfate yielded a soluble and insoluble fraction (CEBRA, 1964; JAQUET and CEBRA, 1965). The soluble component appeared to be divalent because it could form a precipitate with antigen; however, it could not fix complement. This product was similar to the product of complete pepsin hydrolysis obtained by NISONOFF et al. (1960a, 1960b). Recently, L. GOLDSTEIN et al. (1970) synthesized a new carrier and used it to prepare insoluble derivatives of papain, trypsin, and subtilopeptidase A.

b) Myosin and Meromyosin

Water insoluble papain hydrolyzed myosin giving two fractions. One appeared to be identical to light meromyosin and the other was similar to heavy meromyosin (KOMINZ et al., 1965). When bound trypsin was used for the limited hydrolysis of myosin and heavy meromyosin, a highly helical fragment was obtained and its properties were investigated (LOWEY et al., 1966, 1967). The insoluble enzyme's action is different than that of the soluble enzyme because of steric hindrance and electrostatic effects of the polyelectrolyte carrier. The bound enzyme hydrolyzes myosin at a much slower rate and attacks fewer lysine and arginine bonds than the native enzyme, and thus has a modified mode of action.

In further studies on the structure of myosin, a preparation of papain coupled to cellulose was used in a controlled digestion of myosin (NIHEI and KAY, 1968). A component was isolated which possessed three times the ATPase activity of myosin. This material had an average molecular weight of 110000, a sedimentation coefficient of 5.4 S, and one-third the myosin mass. The relationship of this fragment to the parent myosin molecule agreed with earlier data gathered on the structure of myosin (MUELLER, 1965; JONES and PERRY, 1966).

5. Bound Enzymes in the Study of Blood Clotting

STEINBUCH and PEJAUDIER (1962) bound streptokinase to cellulose and used this derivative to produce plasmin from plasminogen, likewise streptokinase was insolubilized by coupling it to a diazotized amino acid copolymer and it retained 2–7% of its activity (A. RIMON et al., 1963). The insolubilized enzyme was easily removed from the reaction mixture, providing a means by which the reaction could be terminated at various stages. This technique was used to separate the activation step from the caseinolytic step in the activation of plasminogen. GUTMAN and RIMON (1964) used the same bound enzyme to reveal that plasminogen and plasmin may be proactivators in the activation sequence, since they form an activator in the presence of streptokinase.

S. RIMON et al. (1966) prepared an insoluble derivative of plasminogen which produced an activator which was not plasmin, when streptokinase was present. Thus, this preparation could serve as a proactivator and was used as such in several studies. The reaction of proactivator and streptokinase to form activator was found to be free of pH dependencies in the 4–10 range, and the conversion of plasminogen to plasmin was optimal in the pH 7.5–9.0 range. Unlike free plasminogen, the bound preparation could not be converted to plasmin by streptokinase, yet it could still serve as a proactivator. This is an example of selective modification of biological activities which may occur on binding an enzyme to a carrier.

HUSSAIN and NEWCOMB (1964) coupled thrombin to a diazotized copolymer of p-aminophenylalanine and L-leucine, giving an enzymatically active derivative used to investigate the intermediate steps in blood clotting. The preparation reacted with fibrinogen to start the blood clotting process. Other investigators (ENGEL and ALEXANDER, 1965; ENGEL et al., 1966) prepared a bound thrombin derivative which could activate zymogens such as plasminogen. In addition, thrombin was bound to cellulose and used to activate bovine factor V (NEWCOMB and HOSHIDA, 1965). ALEXANDER and ENGEL (1970) have recently reviewed the work on insoluble thrombin.

Certain investigators (ALEXANDER et al., 1965, 1966; ALEXANDER, 1966), studying the early steps of the reaction between fibrinogen and bound trypsin, found that cleavage of about eight peptide bonds per fibrinogen molecule was sufficient to prevent coagulation by thrombin. Scission of one or two bonds prolonged the rate of clotting. However, bound trypsin may, under certain conditions, cleave the same linkages of fibrinogen cleaved by thrombin and thus enable it to clot instantaneously. The hydrolysis of 2–8 amide bonds by insoluble trypsin did not change the molecular weight or overall shape of the fibrinogen molecule. Two different trypsin carriers were used in these studies and some differences were noted in their action on fibrinogen (ALEXANDER et al., 1965, 1966).

There are other examples in the literature where the carrier was able to affect the biological activity of the bound protein. For example, prothrombin disappeared and thrombin appeared when prothrombin was treated with polytyrosyltrypsin coupled to a diazotized copolymer of p-aminophenylalanine and leucine (IPTT) (A. RIMON et al., 1966). In addition, factors VII and X were also activated. However, trypsin insolubilized on an ethylene-maleic anhydride copolymer (IMET) digested prothrombin, but no thrombin could be detected. Also, factor X was activated by IMET, but factor VII was not. These effects were attributed to the carriers for the enzymes.

6. Analytical Uses

Bound enzymes show a great potential in the analytical field. These materials retain the specificity of the enzyme and can be used repeatedly. They are amenable to automation and, because they are insoluble, the products of the reaction are easily separated from the enzyme.

RIESEL and KATCHALSKI (1964) bound urease to a diazotized copolymer of p-aminophenylalanine and L-leucine. They packed this material into a small column and used it to analyze serum and urine samples for urea, obtaining values which agreed with those obtained by a standard clinical method. HICKS and UPDIKE (1966) entrapped glucose oxidase and lactic dehydrogenase in polyacrylamide gels. A column of entrapped glucose oxidase was used in conjunction with an oxygen electrode for the continuous assay of glucose solutions (UPDIKE and HICKS, 1967b). Horseradish peroxidase was bound to carboxymethylcellulose and used to assay hydrogen peroxide in solutions which were 1.15 M to 1.15×10^{-6} M (WHEETALL and WELIKY, 1966). Cholinesterase was immobilized in a starch gel (BAUMAN et al., 1965) and used with certain fluorogenic esters in a continuous assay of anticholinesterase compounds, such as certain organophosphorus insecticides (GUILBAULT and KRAMER, 1965).

Recently, glucose oxidase was bound to the inner surface of a chemically modified polystyrene coil and used in a flow circuit for automated determinations of glucose (HORNBY et al., 1970). This tube maintained activity for six weeks. Similarly, urease has been attached to nylon tubes (SUNDARAM and HORNBY, 1970).

An unique method employing bound enzymes for the determination of glucose was developed by UPDIKE and HICKS (1967a). They coated an oxygen electrode with a gel containing glucose oxidase. When this electrode contacted blood or plasma, the enzymic reaction in the gel decreased the amount of oxygen in the vicinity of the electrode and affected its current output. The response of the electrode was proportional to the oxygen consumed and, in turn, to the glucose concentration of the sample. Of course, the oxygen concentration in such a system cannot be rate limiting and the substrate concentration must be well below the Km of the enzyme.

This idea was extended by GUILBAULT and MONTALVO (1969, 1970) to an electrode for urea determinations. Here a cation electrode was coated with polyacrylamide gel containing urease. When this electrode was placed in a solution of urea, the enzymatic reaction produced ammonium ions which were sensed by the electrode. Within certain concentration limits, the potential of the electrode was related to the urea concentration in the sample. A similar technique was used to prepare an electrode specific for L-amino acids (Guilbault and Hrabankova, 1970a, 1970b). A cation electrode was coated with a layer of L-amino acid oxidase. The electrode in contact with an L-amino acid solution senses the ammonium ion produced by the enzymatic reaction. Recently, an enzyme electrode responsive to amygdalin has also been reported (Reichnitz and Llenado, 1971).

7. Industrial Uses

Industrial acceptance of large scale enzymatic conversions has been hindered by the cost of enzymes and the difficulty encountered in recovery, reuse, and separation of the enzyme from the reaction mixture. Insolubilization of the enzyme may answer many of the objections.

The Japanese probably have applied insoluble enzymes in industrial situations to a greater degree than others (Tosa et al., 1966a, 1966b, 1967a, 1967b, 1969a, 1969b). Tosa et al. (1966a) ionically bound amino acid acylases to Sephadex and cellulose in order to prepare columns which could continually separate optical isomers of acyl-DL-amino acids. Using the proper amino acid acylase, they resolved acetyl-DL-methionine into L-methionine and acetyl-D-methionine (Tosa et al., 1966b). Later, these investigators developed an equation to describe the characteristics of this type of column (Tosa et al., 1969b). In addition, Tosa et al. (1969b) studied both DEAE-cellulose and Sephadex as support materials. They found the amount of L-methionine obtained by the column method superior to that obtained from a batch method using native enzyme. The properties of the insoluble DEAE-cellulose- and Sephadex-acylases were similar to those of the soluble enzyme, except for a slight shift in the pH optimum and small increases in heat stability, optimum temperature and resistance to proteolytic enzymes (Tosa et al., 1967b, 1969a). Interestingly, one of the DEAE-Sephadex preparations they obtained was approximately 134% as active as the soluble enzyme under the assay conditions (Tosa et al., 1967b). In addition, an L-amino acid oxidase bound to glass beads was used by column or batch methods to continuously separate a D-amino acid from a DL-mixture (Weetall and Baum, 1969).

There are several examples of the use of bound enzymes in carbohydrate conversions that may have industrial application (Suzuki et al., 1966, 1967; Tsumura and Ishikawa, 1967; Barker et al., 1968, 1969; Wilson and Lilly, 1969). An insoluble saccharase preparation has been packed in a column and used for the continuous inversion of sucrose (Negoro, 1970). DEAE-cellulose was used as a carrier for yeast invertase, and this derivative displayed approximately 50% of the activity of the soluble enzyme in the hydrolysis of sucrose (Suzuki et al., 1967). Barker et al. (1968, 1969) coupled α-, β- and γ-amylase to microcrystalline cellulose and obtained preparations which varied in activity from 6% for bound α-amylase to 16–25% for the insoluble β- and γ-amylases, as compared to the soluble enzymes. Similar results were obtained with preparations of α- and β-amylase coupled to polyacrylamide derivatives

(BARKER et al., 1970b). The recovery and reuse of some of these derivatives have recently been studied (BARKER et al., 1970a).

WILSON and LILLY (1969) used amyloglucosidase covalently coupled to DEAE-cellulose to convert dilute solutions of maltose and dextrin to a product enriched in glucose. Similar results were obtained with finished and unfinished glucose liquors. This bound enzyme was studied for use from an industrial viewpoint, i.e. it was viewed as a chemical reactor in the form of a packed bed. There are a number of other studies which examined the kinetic characteristics of bound enzymes from a similar standpoint (LILLY et al., 1966, 1968; WILSON et al., 1968a, 1968b; LILLY and SHARP, 1968; SELF et al., 1969). Glucoamylase ionically bound to DEAE-cellulose has been used to convert a pretreated starch preparation to a solution containing 5 % reducing sugar (BACHLER et al., 1970). The pH and temperature optima of the bound enzyme were shifted compared to the free enzyme, and columns of the bound enzyme were operated continuously for over three weeks with no loss in enzyme activity.

Penicillin amidase was coupled to cellulose derivatives and used to produce 6-aminopenicillanic acid from benzyl penicillin (SELF et al., 1969), suggesting the use of insoluble enzymes to perform chemical conversions with antibiotics, steroids or alkaloids. In addition, bound enzymes can be used to prepare biochemicals. For example, a column of NAD pyrophosphorylase adsorbed on hydroxylapatite was used to synthesize NAD (TRAUB et al., 1969). A column of adsorbed ATP deaminase was used to prepare IMP from AMP (CHUNG et al., 1968). Soluble enzymes have been recovered from solution through the formation of insoluble precipitates (NEGORO and FUKUMOTO, 1960), and the possibility of using insoluble proteolytic enzymes to clot milk has recently been investigated (M. L. GREEN and CRUTCHFIELD, 1969; JANSEN and OLSON, 1969).

8. Therapeutic Uses

CHANG and his coworkers utilized an interesting method, microencapsulation, for insolubilizing enzymes. They found that they could prepare microcapsules containing enzyme by dispersing the enzyme solution in an organic solvent and then forming a membrane around each droplet. In this manner, cells, cell fragments, cell hemolysates, and enzymes such as carbonic anhydrase, urease, trypsin, uricase and catalase could be enclosed by the membrane (CHANG, 1964, 1966, 1967; CHANG et al., 1966, 1967). These materials may serve as cell models (CHANG et al., 1966; CHANG, 1967), and in the treatment of diseases (CHANG and POZNANSKY, 1968; CHANG, 1971). For example, CHANG and POZNANSKY (1968) showed that mice deficient in catalase died when treated with sodium perborate, yet similar mice treated by injections or with extracorporeal shunts containing microencapsulated catalase were able to detoxify the perborate. Thus it may be possible to treat enzyme deficiency diseases with these encapsulated enzymes. Materials of this nature may also be used to remove toxins from the blood (CHANG, 1966; SPARKS et al., 1969), and as model systems for biochemical studies. Recently, microencapsulated L-asparaginase was used to treat mice implanted with 6C3HED lymphosarcoma cells (CHANG, 1971). These studies showed that the microencapsulated enzyme was more effective than even the free enzyme in retarding the appearance of the lymphosarcoma.

Renin, coupled to an agarose polymer was used to convert angiotensinogen to a vasopressor similar to angiotensin I (SEKI et al., 1970). This material was then converted in a rat to angiotensin II, as evidenced by a blood pressure rise. Thus, bound enzyme preparations may have importance in simplifying the study of systems which depend on the interaction of various factors, since they can be added or removed at any time.

9. Miscellaneous Uses

There have been a number of uses for bound biopolymers which do not fit neatly into any one of the previous categories. However, many of these are interesting applications and deserve attention.

Ethylene-maleic anhydride copolymers were treated with trypsin, α-chymotrypsin and kallikrein to form bound enzyme derivatives. These materials were then employed to isolate pancreatic protease inhibitors (H. FRITZ et al., 1967). The same polyelectrolyte carrier was modified by the introduction of basic groups (H. FRITZ et al., 1969b) to produce a resin capable of isolating inhibitors whose isoelectric points were above pH 4–5, which the untreated copolymer could not do. The modified carriers also made release of the inhibitor possible under more gentle conditions than with the unmodified carrier. The trypsin-kallikrein inhibitor was bound to the modified resin and used to isolate plasmin from bovine serum, and kallikreins from a number of sources (H. FRITZ et al., 1969a). More recently, the isolation of various proteinase inhibitors from body fluids and organ extracts has been described using similar procedures (H. FRITZ et al., 1968; H. FRITZ et al., 1970).

Antigens, antibodies, enzymes, and enzyme inhibitors were polymerized and insolubilized by glutaraldehyde, packed into columns, and used to isolate antibodies, antigens, enzyme inhibitors and enzymes respectively (AVRAMEAS et al., 1969a). Similarly, antigens or antibodies entrapped in polyacrylamide gels could be used to remove their counterpart from a mixture of antigens or antibodies (BERNFELD and WAN, 1963).

Insoluble proteases have been used to activate zymogens (I. SILMAN and KATCHALSKI, 1966; LEVIN et al., 1964). These insoluble proteases are easily removed from the reaction mixture after activation, avoiding interferences caused by the activating enzyme. A bound trypsin derivative was used to activate chymotrypsinogen (LEVIN et al., 1964). Bound thrombin activated chymotrypsinogen A and displayed a higher specific activity toward the zymogen than the soluble enzyme (ENGEL and ALEXANDER, 1966). When pepsinogen was treated with insoluble trypsin, the pepsin formed was different in specificity than the native enzyme. The bound preparation hydrolyzed no more than ten bonds in the zymogen compared to 15 for the soluble enzyme, and different peptide maps were obtained from the digest (ONG et al., 1966). Insoluble protease derivatives have also been used to reduce the immunogenicity and antigenicity of aminopencillanic acid by digesting protein impurities present in the preparations (SHALTIEL et al., 1970).

Nitrocellulose membranes impregnated with proteins have been used to detect substances which interact with the protein layer, e.g. antibodies or enzymes (PRISTOUPIL et al., 1967).

MOSBACH and MOSBACH (1966) entrapped orsellinic acid decarboxylase in polyacrylamide gel and used this material in columns for conversion of substrate. These gel granules showed approximately 30% of the activity of the soluble enzyme. Lichen cells were entrapped in the same polymer matrix and used for similar purposes. Using an entrapped cell to carry out an enzymatic conversion may obviate enzyme purification and stability problems and make enzymatic conversions on an industrial scale more attractive (MOSBACH and MOSBACH, 1966). Spores of *Aspergillus oryzae* were adsorbed on a cellulose derivative and used to enzymatically hydrolyze sucrose (JOHNSON and CIEGLER, 1969). More recently, entrapped fungal cells from *Curvularia lunata* were applied in steroid transformations, as was a Δ^1-dehydrogenase from the bacterium *Corynebacterium simplex* (MOSBACH and LARSSON, 1970). These materials were used in the transformation of Reichstein Compound S to cortisol, and cortisol to prednisolone, respectively. The protozoan, *Tetrahymena pyriformis,* and the bacterium, *Escherichia coli,* have been successfully entrapped in a polyacrylamide matrix (UPDIKE et al., 1969). Both microorganisms appear to remain viable after immobilization in the gel. ROHRBACK and SCOTT (1968) were issued a patent on an electrode infected with microorganisms or enzymes, which is used with a reference electrode to detect any substance which inhibits the enzymes in the cells.

III. Bound Immunochemicals

Bound immunochemicals have been extensively employed in the isolation, purification, and detection of antigens and antibodies. No attempt has been made to completely cover the literature but rather to present examples of the ways in which these materials are used as immunosorbents, as immune indicators or test systems for detection of antigen or antibody, and as substances anchored to increase their immunogenicity. There were, in addition, a few studies of bound immunochemicals not exactly fitting any of these categories.

Studies of bound antibodies have been separated from those of bound antigens in both the immunosorbent and the immunoindicator groups. In some studies, both antigens and antibodies were anchored, and these are usually listed only in the anchored antibody sections.

Recent reviews covered earlier work with bound immunochemicals. SEHON (1967) reviewed chromatographic methods for purifying antibody, including both specific and nonspecific adsorptions. OLOVNIKOV (1968b) reviewed a variety of methods for immunochemical analysis, including immunoadsorption.

I. Silman and Katchalski (1966) included a section on immunosorbents in their review.

A. Bound Antigen Immunosorbents

(Immunosorbents for Antibody)

Insolubilization of antigens or antibodies by coupling or other attachment to insoluble materials is similar to the preparation of insoluble enzymes, although immunogenicity may be retained after procedures or molecular alterations which would impair or destroy enzyme activity.

Campbell (1960) received a patent on antigen or antibody coupling to diazobenzylcellulose for use as an immunoadsorbent. Vannier et al. (1965) coupled tyramine to CM-cellulose using a carbodiimide, and then coupled diazotized p-arsanilic acid to the tyramine. This adsorbent was used to isolate antibodies to the benzenearsonic group, binding 5–10 mg of antibody per gram of adsorbent. Behrens et al. (1967) prepared p-aminophenylbutyrylaminoethyl celluloses and treated them with NaOH before diazotization and coupling of antigen. These immunosorbents specifically bound nearly half their weight in antibody, which could be eluted with acid. Hoyer et al. (1968) prepared and used hapten-cellulose immunosorbents to specifically isolate antibodies to neutral, positively and negatively charged haptens.

Gurvich, Olovnikov, Nezlin, and others in Moscow have conducted extensive work on antigens or antibodies coupled to various carriers. I. Silman and Katchalski (1966) reviewed several of the earlier papers. Gurvich and Kapner (1958) described a quantitative determination of antibodies using a paper precipitation reaction. Gurvich et al. (1961) used diphtheria antigens fixed on cellulose for isolation and determination of antibodies. Olovnikov (1964a) described the use of halogenated alkyl groups on cellulose for coupling antigens to give specific immunoadsorbents. Gurvich (1964) used N-(m-nitrobenzyloxymethyl)-pyridinium chloride for coupling antigens to cellulose powder. Drizlikh and Gurvich (1964) used this same coupling agent to fix antigens to cellulose giving an immunoadsorbent for antibody, for soluble antigen-antibody complexes, or even for antigen once antibody was already adsorbed. Kuzovleva and Gurvich (1966) used the pyridinium reagent to prepare a cellulose ether, reduced the nitro group, reprecipitated the derivative to increase its surface area, and then diazotized and coupled antigens.

Kuzovleva et al. (1966) prepared a stable cellulose ether using 4-(β-hydroxyethylsulfonyl)-2-aminoanisole. This was reprecipitated, diazotized, and coupled to antigens giving an immunoadsorbent binding 250–270 mg of antibody per gram. Kotova (1966) directly introduced the sulfate of 4-(β-hydroxyethyl)-2-sulfanilamidoanisole into cellulose, diazotized, and coupled antigen. Kotova and Dalin (1967) also prepared immunoadsorbents from aminocelluloses.

Nezlin (1961) reviewed methods for isolating pure antibodies. His 1960 paper (Nezlin, 1960) had described adsorption onto cellulose of a purified virus antigen, and the use of this immunoadsorbent for isolation of antibodies against influenza virus. Nezlin and Kulpina (1967) fixed antigen or the antigen-antibody complex to cellulose and used these adsorbents for isolation and purification of the immunoglobulins synthesized by cultures of rabbit spleen and lymphatic cells.

SHAINOFF (1968) prepared a cellulose with monofunctional mercury as the active group by treating cellulose with 1-allyloxy-2,3-epoxypropane and then with mercuric acetate. This mercurial derivative readily bound protein antigens that had been thiolated with N-acetylhomocysteine thiolactone.

HAIMOVICH et al. (1967) described applications of an immunoadsorbent prepared by the reaction of a benzylpenicilloyl derivative of rabbit serum albumin and (bromacetyl)-cellulose. HAIMOVICH (1967) also coupled the bromacetylcellulose to human globulin IgG, generating an adsorbent for antibodies to this immunoglobulin. Sela and his associates have made extensive use of antigens attached to bromacetylcellulose for the isolation and purification of antibodies (ROBBINS et al., 1967). UNGAR-WARON et al. (1967) attached their nucleoside-synthetic polypeptide immunogen to this cellulose derivative and isolated the corresponding antibodies from rabbit serum. JATON et al. (1968) used a related immunosorbent to isolate antibody with a high affinity for the 2,4-dinitrophenyl group. HAIMOVICH and SELA (1969a) used antigens bound to bromacetylcellulose in isolating antibodies specific for bacteriophage T4, poly-D-alanyl groups, or penicilloyl groups. HAIMOVICH et al. (1969b) attached a poly-D-alanyl derivative of rabbit serum albumin to the bromacetylcellulose and used this adsorbent for isolation of antibodies reactive with the poly-D-alanyl derivatives of RNase, human serum albumin, and phage T4. The bromacetylcellulose was also used by GALLAGHER and VOSS (1969) as the carrier for a specific adsorbent of chicken antibody to a dinitrophenyl hapten. A quantitative micromethod for determining antibody was described by GILL and BERNARD (1969). Based on a bromacetylcellulose immunosorbent, it was sensitive enough to determine 25 μg of antibody within ±5%. They found it particularly useful in measuring antibody to synthetic polypeptide antigens. To completely exclude anti-Fc from their anti-light chain serum, GREAVES et al. (1969) adsorbed the rabbit antibodies with Fc attached to bromacetylcellulose. GRAF and UHR (1969) used an immunosorbent based on bromacetylcellulose for removal of circulating antibody from rabbits. They induced two antibodies by injecting BSA and phage. Extracorporeal circulation of rabbit blood over BSA-bromacetylcellulose, or exposure of successive blood portions to the adsorbent before they were returned to the rabbit, removed 32–84% of the anti-BSA but only 12–16% of the anti-phage activity. They suggest use of a specific adsorbent in this manner for treatment of diseases in which circulating antibody contributes to the pathology.

Diazotization to p-aminobenzyl (PAB) cellulose was the method used by TIMPL et al. (1967) to prepare an anchored calf collagen. This immunosorbent specifically removed anticollagen from rabbit serum, and the antibodies were then eluted by collagen peptides. In a second paper (TIMPL et al., 1968), they described removal of antibodies reacting with rabbit skin collagen on PAB cellulose diazotized to this collagen. The effluent from this immunosorbent column the exhibited only antibodies against calf collagen. The antibodies adsorbed on the column could be eluted with tryptic peptides of calf collagen. These antibodies reacted with both the rabbit and the calf collagens. Antitissue serum of high purity was prepared by SURINOV et al. (1968) by removing those antibodies which reacted with serum proteins diazotized to PAB-cellulose. MIHAESCU and CILIEVICI (1969) found human serum albumin (HSA) conjugated to PAB-cellulose was 25-fold more effective than HSA conjugated to m-aminobenzyloxymethyl cellulose as an immunosorbent for antibody. The HSA-PAB cellulose adsorbent gave a 50% yield of 93% pure antibody. This antibody separated on DEAE-cellulose into 2 fractions of different electrophoretic mobility but identical antigenicity.

Carboxymethylcellulose was conjugated via carbodiimide to p-aminophenyl-β-lactoside to give the immunosorbent used by GOPALAKRISHNAN and KARUSH (1969) in isolating antibody to the β-lactoside haptenic group. The antibody was isolated from the sera of rabbits immunized with the β-lactoside attached via carbodiimide to poly-L-lysine.

Aminocellulose was diazotized to insulin by MILES and HALES (1968a), giving an immunosorbent containing 105–185 mg of insulin per g of matrix, and capable of adsorbing 660 mg of anti-insulin per g. Iodination of antibodies while they were on the immunosorbent preserved at least 1 site for antigen binding and permitted easy separation of the ^{125}I-insulin from damaging reagents and contaminating materials.

ORLOV et al. (1969) described selective adsorption of protein antigens onto cellulose cyanurate, 3-chloro-2-hydroxypropyl cellulose ether, or a copolymer of cellulose and glycidyl methacrylate. They did not, however, describe the use of these bound antigens for immunosorption of antibody.

Agarose-based immunosorbents were described by GHETIE and SCHELL (1968). They prepared 1-aminonaphthalene-5-sulfonyl agarose and carboxymethyl agarose, then diazotized or treated with N,N′-dicyclohexylcarbodiimide, and coupled antigens. These immunosorbents held 300 mg of antibody per gram. Agarose beads or aminoethyl cellulose linked via CNBr to blood group substance A gave KRISTIANSEN et al. (1969) fully reversible and highly stable immunosorbents. These adsorbents were useful for purification of anti-A phytohemagglutinin from *Vicia cracca* and the anti-A from human serum.

GURVICH et al. (1967) treated Sephadex with N-(m-nitrobenzyloxymethyl) pyridinium chloride or 4-(β-hydroxyethylsulfonyl)-2-aminoanisole sulfate, then diazotized and coupled antigen. These bound 250–350 mg of antibody per gram of immunosorbent. Sepharose was also the base for the immunosorbent used by REJNEK et al. (1969). They used CNBr to attach a light chain fraction of rabbit immunoglobulin, and then applied this immunosorbent to the isolation of goat antibody specific for this light chain. WOFSY and BURR (1969) described the affinity chromatography of antibodies, antigens and haptens using substances linked by CNBr to Sepharose. They found this type of immunosorbent particularly valuable for isolating antihaptens. The entire matrix is polysaccharide, eliminating the protein or hydrophobic component often used, and reducing non-specific binding. This matrix is also resistant to degradation by the proteolytic enzymes of the serum or ascites fluid being processed. The CNBr-linked immunosorbents were also used by RICHARDS et al. (1969) to examine the heterogeneity of antibodies produced against various dinitrophenyl immunogens. CUATRECASAS (1969a) found insulin attached to Sepharose an excellent immunosorbent, completely removing anti-insulin from crude mixtures containing small amounts of the antibody. He also examined the capacity of 3 modified insulin-Sepharoses to displace insulin from its antibody. Acetylinsulin-Sepharose showed the most effective binding but the lowest activity, insulin-lysyl-Sepharose was the most active and insulin-phenylalanyl-Sepharose was of intermediate characteristics.

A modified Sephadex, the isothiocyanatophenoxyhydroxypropyl derivative, was linked to proinsulin by YIP and LOGOTHETOPOULOS (1969) and used to isolate anti-proinsulin from guinea pig serum. This specific antibody was then used to determine proinsulin in calf serum, but it may also be applicable to beef or human serum.

Immunosorbents based on polystyrene have been described by several investigators. OLOVNIKOV (1962) prepared poly(aminostyrene), diazotized it, and attached antigens. These were used for antibody purification with yields of 30% or greater. KENT and SLADE (1965) obtained a patent on polystyrene-based immunosorbents, related to their earlier studies reviewed by I. SILMAN and KATCHALSKI (1966). The patent included polystyrene or polystyrene derivatives with a variety of chemically reactive groups such as diazo, isothiocyanato, and Hg-containing groups. KITAGAWA et al. (1965a, 1965b) described the use of immunosorbent columns prepared from antigens diazotized to poly(aminostyrene). ORESKES (1966) similarly used γ-globulin diazotized to poly(aminostyrene) for fractionation of rheumatoid factor and rabbit antibodies to the human γ-globulin. SHIMAZAKI et al. (1966) diazotized bovine thyroglobulin to poly(aminostyrene) or p-aminobenzyl cellulose and used these immunosorbents in batch or column purification of rabbit antibodies to thryoglobulin. However, preparation of immunoadsorbents by methods employing dodecyl sulfate or mercaptoethanol were unsuccessful since these reagents destroyed the antigenicity of the thyroglobulin. CHUBB (1967) used poly(aminostyrene) coupled to bovine serum albumin or Rous sarcoma virus in attempts to purify chicken antibodies to these antigens. Unfortunately, only 1% of the virus-neutralizing activity was recoverable. Isothiocyanopolystyrene was the support for antigens used in the immunosorbent columns of MAN'KO et al. (1968). They cited results with anchored horse serum albumin in evaluating this adsorbent.

Polymerized rabbit serum albumin was used by PRESSMAN and his associates as the carrier for immunosorbents. Polymerization was by S-acetylmercaptosuccinic anhydride. TAKAHASHI et al. (1968) used bis-diazobenzidine (BDB) to link antigens to the polymer. They used the adsorbent to isolate fluorescein- and rhodaminelabeled antibodies to γ-, α-, or μ-heavy chains of human immunoglobulin. In similar fashion (TAKAHASHI et al., 1969), they isolated rabbit and goat antibodies specific for $\varkappa$ or λ type light chains. FREEDMAN et al. (1968) treated antibodies specific for the p-azobenzenearsonate hapten with maleic anhydride. When 100% of the amino groups were maleylated, the antibodies lost 85–90% of their combining sites. A solid immunosorbent separated these antibody molecules into two types, only one of which reacted with the adsorbent and was retained on it. Mouse antibodies to the p-azobenzoate group were recovered in good yield from ascitic fluid by MATTIOLI et al. (1968) using solid immunosorbents. They noted that hemocyanin was a better carrier for this hapten than was bovine serum albumin.

BAUMINGER et al. (1969b) used genistein linked to a heterologous carrier as an adsorbent for rabbit antibodies to the genistein. MORENO and KABAT (1969) used columns of the insoluble polyleucyl derivative of hog blood group substance A to isolate human antibodies to the A substance. The antibodies were eluted by the reduced form of a pentasaccharide with group A activity. KAPLAN and KABAT (1966) prepared insoluble blood group substances by copolymerizing them with N-carboxyleucine anhydride, and used them as specific adsorbents for antibodies. GIVAS et al. (1968) attached myoglobin to the copolymer of ethylene and maleic anhydride, then mixed this with Sephadex G-25. This was used as an immunosorbent column for the isolation of antibodies.

Bovine hide powder was coupled to haptens by B. LEVINE and LEVYTSKA (1969) and used as the immunosorbent for rabbit antibodies to the haptens. The benzylpenicilloyl and dinitrophenyl hide powders both had high adsorptive capacities for antibody. Hemocyanin was trinitrophenylated and then azocoupled by haptens by BING et al. (1969). Although soluble above pH 7, these conjugates quantitatively precipitated at pH 5–6 and were used as high capacity immunosorbents for antihapten antibodies from rabbit, mouse, and equine sera.

An immunochemical coupling technique was used by MAGE et al. (1969) to prepare an immunosorbent. Antibody to sheep red cells was adsorbed on polyurethane disks and sheep erythrocytes were then attached. These disks were then used to isolate spleen cells with an immunocompetence against sheep erythrocytes, from mice immunized with these erythrocytes.

Cross-linked, insoluble antigens are sometimes useful as immunosorbents. JONIAU et al. (1968) used glutaraldehyde to polymerize proteins for use as immunoadsorbents. TALLBERG (1969) used ethyl chloroformate to polymerize tetanus toxoid. This insoluble toxoid was mixed with Sephadex to prevent clogging of the filter bed, and then used as a high capacity, reusable adsorbent for good yields of human antibody against the toxoid. G. B. GOLDSTEIN et al. (1969) used the ethyl chloroformate to prepare insoluble polymers of gliadin and ragweed antigen E. These insoluble allergens removed antibodies reactive in Prausnitz-Kuestner and tanned cell hemagglutination tests from the sera of persons allergic to wheat or to ragweed. The reagins could be eluted from the immunosorbents by salt solutions of high ionic strength. Cross-linking of thyroglobulin by bis-diazobenzidine was used by TORRIGIANI and ROITT (1969) to prepare an immunosorbent for autoantibodies to the thyroglobulin. WEETALL (1967) polymerized microorganisms and used them as a specific immunosorbent for the isolation of large amounts of antibody to the bacterial antigens.

OLOVNIKOV (1964b) found tetrazotized p-diaminodiphenylamine relatively stable and more easily stored than BDB, and used it to form antigen complexes useful as immunosorbents. DANDLIKER et al. (1968) used carbodiimide coupling agents to interconnect antigen molecules, or to couple antigen to cellulose, CM-cellulose, or aromatic diamine derivatives of CM-cellulose. They defined optimum conditions for these coupling reactions. CARPENTER and BARSALES (1967) used a carbodiimide to couple antigen to bentonite particles. Further study by CARPENTER and REISBERG (1968) suggested an organometallic bond between antigen and bentonite. This adsorbent bound 0.17 mole of antibody per mole of adsorbent antigen.

DONEY et al. (1968) used antigens entrapped in polyacrylamide gel for selective removal of antibodies from crude antisera. The O-acetylated galactan of gum arabic was trapped in a column of cellulose and used by JACKSON et al. (1968) to isolate rabbit antibody reactive with *Salmonella typhimurium* antigen 5. This antibody was also found in normal rat serum, correlating with the natural resistance of the rat to infection by *S. typhimurium*. Trapping of antigen in polyacrylamide cross-linked by N,N′-methylenebisacrylamide gave CARREL et al. (1969) an immunosorbent which they reused as many as 15 times without apparent loss of its capacity to bind antibody to the IgG trapped in the gel. This polyacrylamide carrier was mechanically stable, insoluble in water, relatively resistant to chemical reactions, and showed little or no nonspecific binding of serum proteins.

Antigens coated on beads of plastic or glass were used in columns by WIGZELL and ANDERSSON (1969) to isolate murine immunocompetent cells with receptor sites for the antigens (various albumins). They found evidence for a true receptor site on these cells, not merely an adsorbed antibody.

Dissociation of antibody from immunoprecipitates is a method of antibody isolation and purification related to the immunosorbents. BEAUMONT and DELPLANQUE (1969) used this method to obtain a high yield of rabbit antibody to β-lipoprotein. The antigen-antibody complex was dissociated in an alkaline buffer containing 4 M urea, and the antibody further purified by chromatography on DEAE-cellulose. MUELLER (1966) dissolved the antigen-antibody precipitate in 0.1 N HCl, neutralized the solution with $NaHCO_3$, and obtained pure antibodies in the supernatant after centrifugation. Ragweed antibodies were isolated by SPITZER et al. (1968) from the immune precipitate after its dissociation in glycine-H_2SO_4 buffer of pH 2.7.

B. Bound Antibody Immunosorbents

(Immunosorbents for Antigens)

One of the first immunosorbents for antigens, and the first use of bromoacetyl cellulose as a support for antibody was described by JAGENDORF et al.

(1963). Steinbuch et al. (1965) and Matte et al. (1965) coupled diazotized p-aminobenzyl-cellulose to globulin fractions giving adsorbents for the isolation and study of rheumatoid factor. Galvez (1966) coupled antibodies to diazotized p-aminobenzyl-cellulose, defining the best conditions for the coupling. He used these immunoadsorbents in columns for the specific adsorption of plant viruses.

R. V. Davis et al. (1969) attached antibodies or antigens to diazotized m-aminobenzyloxymethyl cellulose in their preparation of immunosorbents. They cited as examples: anchored BSA as an immunosorbent for rabbit antibody to the bovine serum albumin, and a sorbent for C-reactive protein based on anchored antibody. Similarly, Rokhlin and Nezlin (1969) diazotized antibody to small cellulose particles and used these immunosorbents for determination of antigen. These fixed antibodies adsorbed 2–3 times as much antigen as did antibodies fixed through an antigen-antibody sandwich technique. Gallop et al. (1966) used γ-globulin coupled to aminocellulose for fractionating mixtures of antigens.

Wide, Porath, and others at Uppsala have prepared immunosorbents by coupling proteins to Sephadex or agarose. Wide and Porath (1966) coupled antibodies to isothiocyanatophenoxyhydroxypropyl-Sephadex and applied these insoluble antibodies to assays for several hormones and for immunoglobulin. Wide et al. (1967) compared three cyanogen halide methods and an isothiocyanate method for coupling antibodies to derivatives of Sephadex and favored CNBr-activation because such immunosorbents were easiest to prepare, were stable for at least 3 months, and had high antigen capacity. The paper by Porath (1968) on molecular sieving and adsorption included a discussion on preparation and properties of immunoadsorbents based on Sephadex or agarose. Porath found these materials, particularly agarose, superior to other polymers as carriers of antigens when one attempts to release the antibody. He noted the advantages of using cyanogen bromide as the coupling agent. Wide (1967) presented a paper on the use of Sephadex-coupled antibodies for the radioimmune determination of proteins. Kristiansen and Porath (1968) prepared an immunosorbent column for the purification of blood group substance A. Cyanogen bromide was used by Givol et al. (1970) in attaching antibody to Sepharose. This anchored antibody was specific for the dinitrophenol (DNP) hapten and was used to isolate DNP-peptides.

Catt and his co-workers have published a series of papers on solid-phase radioimmunoassays. Catt et al. (1966, 1967c) attached antibody to an isothiocyanate derivative of the graft copolymer of styrene and polytetrafluorethylene. This immunosorbent prepared with antibody to human growth hormone irreversibly bound the hormone. Binding of unknown or standard amounts of unlabeled hormone followed by radiolabeled hormone was used for hormone assays. Preparation of the immunosorbent in the form of thin wafers with uniform size and surface further simplified the procedure. They presented further technical details (Catt et al., 1967b) for the application of the assay, including simultaneous determination of growth hormone and LH by using immunosorbent disks to which both antibodies were bound. The use of the immunosorbent as plastic disks was stressed in another paper (CATT et al., 1967a), which also mentioned successful radioimmunoassay of fibrinogen. Further details of the disk radioimmunoassay of LH and simultaneous assay of LH and growth hormone were given by CATT et al. (1968b). CATT and TREGEAR (1967) described a modification of these techniques, in which they attached the antibody to disposable tubes of the polymeric carrier.

SUTHERLAND (1960) claimed that careful denaturation with heat or chemicals could coagulate antibodies, generating an immunosorbent, without loss of specific combining sites. McDUFFIE and SUTHERLAND (1963) and SUTHERLAND and McDUFFIE (1963) further described these coagulated antigen-antibody complexes.

AVRAMEAS and TERNYNCK (1966) prepared insoluble but active protein polymers by adding aqueous solutions of proteins to a refrigerated mixture of $CHCl_3$, Et_3N, and ethyl chloroformate. Antibody insolubilized in this manner was used as an adsorbent for the specific preparation of human serum albumin. AVRAMEAS and TERNYNCK (1967) noted that ethyl chloroformate apparently links the protein molecules giving the insoluble polymers useful as specific, efficient, stable immunosorbents.

Antibodies polymerized by ethyl chloroformate were used by TALLBERG et al. (1968a) to purify human chorionic gonadotrophin, human growth hormone, and human placental lactogen. Insulin was removed from human plasma by THORELL (1968) using antibodies coupled to Sephadex. This gave a plasma free of insulin, useful as a diluent in radioimmunoassays.

An antigen specific for human thymus was isolated by TALLBERG et al. (1968b) using polymerized antibody as an immunosorbent. They preferred an adsorbent for contaminating antigens rather than an adsorbent for the thymus antigen, since dissociation from the immunosorbent often destroyed the quaternary structure and biological activity. An antigen specific for human liver was isolated by immunoadsorption in the laboratory of SCHUMACHER and SCHNEIDER (1969). They stressed the need for monospecific antibody in these antigen isolations, using antisera adsorbed with various antigens from other organs to gain this specificity. An antigen specific for Ehrlich ascites (EA) was isolated by SUTHERLAND and DATTAGUPTA (1969) using an adsorbent made by ethanol-coagulation of the immuno-precipitate from the reaction of tumor extract and chicken antiserum to this extract. The chicken antiserum had been previously treated with antigens from normal mouse tissues to increase its specificity for the EA antigen. The EA antigen eluted from the coagulated immunosorbent gave a single line on immunodiffusion against the chicken antiserum from which antibodies to normal tissue antigens had been adsorbed.

Using human immunoglobulins as antigens, TALLBERG and JORMALAINEN (1968) successfully separated IgA, IgM and IgG directly from human serum. Their immunosorbents were monospecific rabbit antisera polymerized by ethyl chloroformate. The Ig fractions obtained from these adsorbents were then used to immunize rabbits. The IgG induced antibody to IgG, with only minute amounts of anti-IgM. The IgM induced strong antibody titers against both

IgM and IgG. The IgA from the sorbent caused the rabbits to produce antibodies against all three Ig groups. RADERMECKER and GOODFRIEND (1969) modified the polymerization by ethyl chloroformate in order to reduce the nonspecific adsorption of other serum proteins. Antibody to IgG was polymerized by this modified procedure and used to successfully remove IgG from allergic serum without appreciably affecting the titer of reaginic antibodies. IgA was isolated by ZSCHOCKE et al. (1969) from pooled human plasma by several fractionation steps, with the last step being removal of IgG on an immunoadsorbent of anti-IgG attached to bromoacetyl cellulose. ITO et al. (1969) used the ethyl chloroformate polymerized antibodies against the various Ig groups. Only the polymerized anti-IgE removed reaginic activity from the sera of allergic patients. Eluates from this adsorbent passively sensitized human skin. Polymerized antibodies to IgA, IgM, IgG and IgD removed none of the reagin.

Rabbit antiserum to normal human serum was polymerized by KARITZKY and KUNZER (1967) using ethyl chloroformate. This polymer removed other blood constituents from a solution of human fibrinogen. The pure fibrinogen was then used to immunize rabbits, giving a rabbit antiserum specific for fibrinogen. Polymerized antifibrinogen was then used (KARITZKY and KUNZER, 1968) to produce human plasma deficient in fibrinogen but not deficient in other clotting factors.

AVRAMEAS and TERNYNCK (1969) discussed the use of glutaraldehyde to cross-link antibodies or protein antigens. These insoluble antibodies or antigens were useful in columns or in batch processes as efficient, stable, and specific adsorbents. Glutaraldehyde-insolubilized antigens or antibodies were further described by AVRAMEAS et al. (1969c).

STEPHEN and SMITH (1965) described the use of disulfide-linked antibodies in Sephadex columns for antigen isolation. STEPHEN et al. (1966) further described preparation of these insoluble polymers by thiolation and then disulfide cross-linking of immune globulins, and use of the poorly soluble polymers as adsorbents for antigens. CHIDLOW et al. (1967) described reactions of a commercial cytochrome c with disulfide-linked antibodies from rabbits. They described further work with disulfide-linked antibodies and the conditions critical for desorption of unaltered antigen (CHIDLOW et al., 1968). An anti-lysozyme immunosorbent was used five times with no appreciable loss of specificity.

Immunosorbents for virus have been prepared. WOOD et al. (1968) used disulfide-linked antiserum from rabbits hyperimmunized with Semliki Forest virus. Thiolation and oxidation cross-linked the γ-globulin concentrate, giving an insoluble polymer which effectively adsorbed the virus. Desorption yielded a pure virus preparation with little loss of infectivity. AVRAMEAS et al. (1969a) polymerized antibody with ethyl chloroformate or glutaraldehyde to produce immunosorbents for rat K virus or polio virus. They also suggest the use of immunosorbents specific for contaminating materials when fragile viruses are to be isolated, since the adsorption-desorption of the virus itself might damage it.

Foot-and-mouth disease viruses were isolated and differentiated into types by MATHEKA and MUSSGAY (1969) using immunosorbents prepared by hooking antibody onto activated Sepharose.

EVANS et al. (1969) described the use of antibody coated onto polyurethane foam for the specific removal of cells from a suspension. Adsorption of red cells was described, and the cells could then be readily recovered from polyurethane disks.

WOLF and IZENBERG (1969) thiolated antibody before using it to precipitate its specific antigen. The precipitate was then collected, dissociated at pH 1.8, and the antibody cross-linked with 3,6-bis(acetoxymercurimethyl)-dioxane. This precipitated the antibody leaving a purified antigen free of antibody. Similarly, EMMART and BATES (1968) purified ovine prolactin, ovine growth hormone, and fish prolactin by precipitating them with specific antibody, dissociating the antigen-antibody complex at low acidity, and separating the antigen and antibody by Sephadex chromatography.

C. Bound Antibodies as Immuno-indicators

(Immunoindicators for Antigens)

OLOVNIKOV (1968a) reviewed methods for determining antigen based on agglutination of polymerized antibody, or of this polymer attached to erythrocytes. YDE (1969) reviewed the several techniques that have been used for the radio-immunochemical determination of human plasma growth hormone (HGH). Most of the recent modifications have been improved methods for separating the free HGH from that bound by antibody, including anchored antibody for easy removal of the bound form.

GOTTLIEB et al. (1967) adsorbed γ-globulin on small particles of charcoal and used these particles for detection of rheumatoid factor. They claim that appropriate charcoals or carbon blacks almost instantly adsorb a wide variety of antigens or antibodies and make stable, inexpensive, ready-to-use carriers to substitute for latex, bentonite or erythrocyte particles.

Cellulose-bound antibody was used by HABERMANN and RAKER (1969) and by HABERMANN et al. (1969) for determination of staphylococcal α-toxin. They used bromacetyl cellulose for covalent binding of the antibody molecules. This same carrier was used by MANN et al. (1969) for covalent binding of antibodies specific for each of the types of human immunoglobulins. These insoluble antibodies were applied to the radioimmune determination of the levels of the various immunoglobulins.

HAIMOVICH and SELA (1969b) described the covalent attachment of antibody or antigen to phage, giving an immunoindicator which detected antigens or antibodies by their inactivation of the phage. An inactivation inhibition test was also possible, in a fashion analogous to the widely-used hemagglutination inhibition (HAI) tests. The HAI technique was applied by NYERGES (1969) to the determination of antivaccinia titers, using a bovine antivaccinia globulin attached to tanned sheep erythrocytes. Agglutination of these cells by vaccinia virus could then be inhibited to a degree corresponding to the amount of antivaccinia activity in the human serum added to the test chamber.

There were two recent papers of interest to those employing immune hemolysis as an indicator of antigen-antibody reactions. INAI and TSUYUGUCHI (1968) noted a complement-mediated hemolysis of sheep erythrocytes in the absence of an antigen-antibody reaction. Aggregated human IgG was coupled via $CrCl_3$ to the erythrocytes, and, in guinea pig serum, these erythrocytes reacted with complement in much the same manner as erythrocytes optimally sensitized by antibody. Erythrocytes coupled to IgM were also lysed by complement, but not those coupled to IgA. Even simple adsorption of a complement-binding protein was sufficient to induce lysis by complement. Similarly, attachment of the C′1 component of complement by BDB, $CrCl_3$, or tannic acid treatment gave LINSCOTT et al. (1969) red cells which were lysed by whole guinea pig complement in the absence of antibody.

A hybrid antibody was prepared by HAMMERLING et al. (1968, 1969) and used for the location of cellular antigens by electron microscopy. Rabbit antiferritin and rabbit anti-mouse Ig were treated with pepsin and dissociated into fragments which could be dimerized to give an antibody molecule with sites for both ferritin and the mouse globulin. This hybrid antibody could attach ferritin to a mouse antibody that had been used to locate the cell surface antigens on leukemia cells. Ferritin-labeled antibodies were used by HAFERKAMP et al. (1969) for the electron microscopic investigation of erythrocyte agglutination. They found agglutination between cells sensitized by antibody and freshly added, non-sensitized cells without antibody transfer to these fresh cells. A sandwich label was used in part of this study, attaching rabbit antibody to sheep cells and then ferritin-labeled sheep antibody to the rabbit globulin.

Fluorescein-labeled antibody was used by NAYAK et al. (1969a, 1969b) to examine cellular locations of growth hormone and the nature of the antigenic sites on this hormone. They found growth hormone only in the acidophils of the pituitary gland in the species studied. Multiple antigenic sites were indicated on the growth hormone.

Antibody polymerized by ethyl chloroformate was used by TALLBERG and his associates for detection and determination of antigens, much in the manner they used it as an immunosorbent for antigen isolation and purification. TALLBERG and LINTURI (1968) used small particles of polymerized guinea pig anti-insulin in their radioimmune assay for insulin. TALLBERG and ESKOLA (1969) determined serum levels of human placental lactogen (HPL) using polymerized-antibody particles coated with antigen. Agglutination of these particles by anti-HPL was then inhibited by the HPL in the patient's serum. TALLBERG et al. (1969) summarized results of these 2 applications and described an assay for human chorionic gonadotropin similar to that for HPL. TALLBERG (1967) used ultrasonics to reduce the size of the particles of polymers to which antigens or antibodies were bound. These smaller particles were more sensitive than were the larger particles in slide agglutinations.

Trapping of antibody in polyacrylamide gel gave GOODFRIEND et al. (1969) a solid reagent for the radioimmunoassay of angiotensin. They suggest general application of this technique, since the antibody is trapped without chemical or physical denaturation.

QUABBE (1969) used centrifugation rather than filtration in a modification of the Hales and Randle radioimmunassay for insulin. He included precipitation of some other antigen-antibody complex, to give a carrier immunoprecipitate and better separation of the bound insulin.

Adsorption of antibody on bentonite was used to separate bound FSH in the radioimmunassay for this hormone devised by BUTT and LYNCH (1968). This bentonite adsorption of antibody was also used by CHENG and TALMAGE (1969) in radioimmune determinations of antigens. They also made applications of this adsorption to the detection of antibodies.

KING (1968) studied protein adsorption onto guinea pig tissue slices. Adsorbed antibody could still react with antigen, indicating that the adsorbed protein was on the surface of the tissues.

Inhibition of immune cell migration in the presence of antigen was observed by CARPENTER et al. (1968). Spleen, lung, lymph node, and peritoneal cells readily migrated in a simple agar medium, but this migration was blocked by antigens used to immunize the animal donors of these cells. Antigen also blocked migration of peritoneal cells in capillary tubes.

Antibody adsorbed to test tubes was the most sensitive and convenient method among those used by GOODFRIEND et al. (1968) for radioimmunassay of angiotensin II. ASKENASE and LEONARD (1969) used antibody adsorbed to polystyrene tubes in their radioimmune determination of β1c-globulin. Immunochemical sandwiches were built by SALMON et al. (1969) upon disposable plastic tubes or isothiocyanated plastic disks. These solid phase immunochemicals were then applied to the radioimmune determination of immunoglobulin components. By first attaching antigen to the plastic, then antibody, and measuring the attachment of labeled antigen in the presence and absence of the antigen being assayed, they found enhanced antigen binding and greater sensitivity than in techniques where the antibody was coated directly on the plastic. Antiserum coated on the walls of plastic tubes was used by ABRAHAM (1969) for radioimmunassay of 17β-estradiol. CATT et al. (1968a) coated antibody on disks of poly-(tetrafluoroethylene-p-isothiocyanatostyrene) and developed a radioimmunassay for fibrinogen and products of the action of plasmin on the fibrinogen.

DEAE-cellulose paper collected the antigen-antibody complex in the filtration technique used by GRANT and SIMON (1968) for determination of antigens. They used purified antibodies labeled with ^{125}I in assays for bovine serum albumin, the B protein of the tryptophan synthetase of *E. coli*, and *B. subtilis* flagella. The DEAE-cellulose also trapped the free antigens, but the unbound antibody passed through.

Immunoelectroadsorption has been extensively investigated at Rockefeller University. The fundamentals of the method were recently described by

ROTHEN and MATHOT (1969b), based on a review of their work since 1964. Since their review lists the earlier references, we have included only their papers since 1968. The assay is based on adsorption of a layer of antigen and then a layer of antibodies onto a metalized slide. Both adsorptions are enhanced by application of a small electric current of the proper polarity. The thickness of the layer is a function of the concentration of antibody. Some of the physicochemical factors were discussed by MATHOT and ROTHEN (1969b, 1969c). At the 158th ACS meeting, ROTHEN and MATHOT (1969a) described results with type III polysaccharide from the pneumococcus. Immunoelectroadsorption on a slide treated first with a solution containing only 10^{-13} g of polysaccharide per ml, then with antibody, gave an antibody layer at least 10 Angstroms thicker than on slides without the specific antigen. Similar sensitivity for antibody detection could be obtained by coating the slide with more of the polysaccharide, permitting detection of a potent rabbit antiserum diluted 1 to 400000. ROTHEN et al. (1969) described application of the method to the determination of human or bovine growth hormone, with a sensitivity of 0.1–0.2 ng/ml.

The observations of HIRATA et al. (1969) are somewhat related. They noted absurdly high antiserum titers using the Takatsy stainless steel loops for serial dilutions of the rabbit antiserum to bovine serum albumin. They traced this discrepancy to adsorption of antibody onto the metal loop, then its gradual release during the dilution procedure. Loops presoaked in a gelatin and rabbit albumin solution did not have this carry-over effect. Treatment of antiserum with zinc, iron or tin granules diminished the carry-over.

An interesting electrical measure of the antigen-antibody reaction was described by BARFORT et al. (1968). The electrical resistance of a bimolecular lipid membrane with antibody and complement on one side decreased when antigen was added to the other side. They described determinations using only 0.1 μl of antiserum.

D. Bound Antigens as Immuno-indicators

(Immunoindicators for Antibody)

Much of the work on these immunoindicators has centered on the use of the red blood cell as the carrier particle. Various antigens or haptens were attached by coupling agents, or simply by adsorption to tanned cells or to aldehyde-treated cells. Other carriers have been used, such as phage, synthetic polymers, bacterial cells, or large protein molecules. In many cases the complete test was for antigen by its inhibition of the reaction with antibody.

Patents. The test for human chorionic gonadotropin (HCG) patented by MATUHASI and KOBAYASHI (1969a, 1969b) used the reaction of HCG adsorbed on tanned sheep erythrocytes with rabbit anti-HCG. Agglutination, or its inhibition by urine containing HCG, was determined in tilted capillary tubes. Agglutination produced a coarse velvet precipitate on the inner wall of the tube, while lack of agglutination gave a definite line of cells. Thus the immunoindicator for antibody (the HCG-erythrocyte) is used as a test for antigen by virtue of the hemagglutination inhibition by this antigen (HCG).

HIRATA (1969) patented processes for preparing stabilized erythrocytes and increasing the hemagglutination titers after antigen coating. Erythrocytes from various animals were treated with small amounts of pyruvaldehyde, then with small amounts of formaldehyde. These double aldehyde-stabilized cells were coated with fibrinogen prior to coating them with antigen.

The French patent of ARQUILLA (1968) described the use of microorganisms as the carrier particles for antigens attached by a variety of coupling agents. These antigen-coated organisms reacted with antibody to produce agglutination on plates, slides, or in tubes, or agglutination detectable by chromatographic techniques.

The BOZICEVICH (1969) patent described a test for rubella antibodies using viral antigen adsorbed on bentonite. Agglutination of these colloidal particles by antibody was viewed under low magnification inside a wax ring on the slide.

BENNICH et al. (1969) patented the radioimmune detection of reagins. Allergens were chemically coupled to an insoluble polymer (dextran or its derivatives have been widely used by this group), contacted with the patient's serum and then with radiolabeled antibody against the reagin immunoglobulin. Alternatively, the allergen-polymer was contacted with the patient's serum, then with unlabeled anti-reagin, then with labeled reagin. They included descriptions of the preparation of reagin immunoglobulin and its labeling, and of antibodies against this immunoglobulin.

Papers. Chemical coupling of antigens and red blood cells received some study and considerable application in recent investigations. HABER and ROSE (1968) established optimum conditions for coupling by $CrCl_3$, bis-diazobenzidine (BDB), and tolylene-2,4-diisocyanate, and for adsorption of antigens by tanned cells. These four techniques for attaching bovine serum albumin gave equal titers with anti-BSA. The tanned cell method gave the most sensitive test for antibody to rabbit thyroid extract. Coupling with BDB gave the most sensitive indicator for anti-insulin. AVRAMEAS et al. (1969b) used glutaraldehyde, cyanuric chloride or tetrazotized o-dianisidine to couple antigens to erythrocytes. These indicators were at least as sensitive as antigen-erythrocyte complexes made with BDB or tanned cells. These investigators are strong proponents of the glutaraldehyde couple, and noted that such cells retained their sensitivity after lyophilization.

GOLD and FUDENBERG (1967) found $CrCl_3$ coupled antigens to erythrocytes without destroying the specificity of the antigen. Papain treatment of the $CrCl_3$-treated cells, or use of bovine serum albumin rather than saline as the diluent, increased antiserum titers. GOLD (1968) further described the effects of the $CrCl_3$ treatment. RENOUX et al. (1968) successfully coupled a glycoprotein from *Brucella abortus* to erythrocytes with $CrCl_3$, but the indicator attempted by a BDB couple was not agglutinated by antibody. DUBOIS et al. (1968) found that BDB failed to couple the glycoprotein but $CrCl_3$ produced an effective indicator in an immune hemolysis system. Both BDB and $CrCl_3$

effectively coupled prolactin to the erythrocytes, giving immune hemolysis systems of equal sensitivity for detecting anti-prolactin. Dubois (1969b) then described the use of both the BDB and $CrCl_3$ coupling of prolactin in systems where inhibition of immune hemolysis was used for the determination of prolactin. The effects of volumes and concentrations of materials during the $CrCl_3$ coupling were examined by Dubois (1969a), using the prolactin immune hemolysis system. Renoux and Renoux (1969) described the use of immune hemolysis in gels for the detection of spleen cells producing antibody to bacterial antigens. Splenic immunocytes, from mice immunized with bacteria, produced hemolytic plaques in agar containing erythrocytes coupled via $CrCl_3$ to the bacterial antigens.

$CrCl_3$ was also used as a coupling agent by Vyas et al. (1968). They coupled various γ-globulins and myeloma proteins to human group 0 erythrocytes and used hemagglutination tests for typing the Gm and Inv characteristics of human sera. Perucca et al. (1969) used $CrCl_3$ to couple ovalbumin to erythrocytes. The sensitivity of this model system for passive immune hemolysis was compared to passive cutaneous anaphylaxis and hemagglutination for the determination of antibody titers. $CrCl_3$ was used by Baker et al. (1969) to attach purified pneumococcal polysaccharides to erythrocytes. These sensitized erythrocytes could be used for antibody determination by hemolysis or hemagglutination tests or were incorporated in agar gel for detection of antibody-producing cells by the hemolytic plaque technique.

Weinbach (1969) fixed the erythrocytes of various animal species by treatment with glutaraldehyde, then attached antigens by $CrCl_3$ or by merely changing the ionic strength of the medium. These were used in passive hemagglutination tests. The glutaraldehyde-fixed cells, by themselves, were useful in direct or viral hemagglutination tests.

W. C. Davis et al. (1968) used the $CrCl_3$-hemagglutination indicator to assay antibody to ferritin-labeled fractions of γ-globulin. Application of ferritin-labeled antibody in electron microscopy suggested discrete sites for the binding of exogenous antigens to erythrocytes, perhaps related to undiscovered determinants on the red cell.

A carbodiimide was used by Dietrich and Frischknecht (1968) to couple hypertensin to sheep erythrocytes. Inhibition of hemagglutination by related peptides helped reveal the antigenic determinants of this synthetic angiotensin II amide. Carbodiimide-coupling of pregnant mare serum gonadotropin (PMSG) to aldehyde-fixed sheep erythrocytes gave Hoff and Cohen (1969) an indicator they could use with rabbit anti-PMSG to assay PMSG by a hemagglutination-inhibition technique. This assay showed no cross-reactions with human chorionic or menopausal gonadotropins, nor with ovine LH, but reacted well with ovine FSH. The indicator was still reactive after storage at 4° for 1 year.

Glutaraldehyde was used by Onkelinx et al. (1969) to couple various antigens to erythrocytes. These indicators had sensitivity comparable to that obtained by BDB-coupling. Tribouley et al. (1969) used glutaraldehyde to couple sheep erythrocytes to a delipidated antigenic extract of *Fasciola hepatica*. Freezing of the indicator did not lower its sensitivity for the determination of rat antibodies, indicating the stability of the glutaraldehyde attachment of antigen.

Diazotized p-aminophenylmercuric acetate was used by AMOS (1968) as the link between human group 0 erythrocytes and penicillamine. This was a hemagglutination indicator of hypersensitivity to penicillamine, but gave titers of only 1 to 32 with positive sera. GYENES and SEHON (1964) used tolylene 2,4-diisocyanate to couple antigens to erythrocytes. A. RIMON and SELA (1966) prepared peptidylated erythrocytes by their reaction with N-carboxyaminoacid anhydrides and found them useful for detecting antibodies to the attached peptides.

Red cells with phage attached via BDB were used by VICARI et al. (1968) for hemolytic plaque determinations of rabbit spleen cells producing antibody against the phage ΦX174. HIRATA and STASHAK (1965) used antigens adsorbed or BDB-coupled to erythrocytes in agglutination tests for the attached antigens.

A double aldehyde treatment was used by HIRATA and STALL (1969a) to fix rabbit erythrocytes first with pyruvaldehyde then with formaldehyde. After coating with bovine serum albumin (BSA) at low pH, they were repeatedly frozen and thawed to increase the sensitivity for detection of anti-BSA. The indicator was stored in liquid nitrogen since it gradually deteriorated at −20°. HIRATA and STALL (1969b) also found that periodate oxidation or exposure to low pH increased the hemagglutinability of the indicator. The effects of low pH and freeze-thawing were additive. HIRATA and BRANDRISS (1968) had earlier described the usefulness of indicators in which the red cells had been stabilized by formaldehyde or pyruvaldehyde used singly. HAMMOND et al. (1968) sensitized the formalinized but untanned red cell with carbohydrate antigens of *Neisseria meningitidis*, but only after removal of sialic acid from the partially purified antigens.

The tanned red cell was examined by WOLBERG et al. (1969b) for adsorption of bovine serum albumin in generating a hemagglutination indicator. Mildly denatured BSA was more effectively adsorbed and this increased the sensitivity of the indicator. They described (WOLBERG et al., 1969a) the conditions of heat denaturation which gave antigen best adsorbed and the nature of the denatured fraction showing this increased sensitization of the erythrocytes. WALKER et al. (1969) found tanned erythrocytes effectively adsorbed phage T2 if the phage were first treated with sodium dodecyl sulfate. This hemagglutination indicator for antibody to the phage was at least as sensitive as the phage neutralization test. WALKER and FISHMAN (1969) applied the hemagglutination test to determination of anti-phage antibody produced by cultures of lymph node cells stimulated by an RNA taken from macrophages previously in contact with the tail fibers of phage T2.

The tanned, formalinized red cell was used by HAWKINS (1969) for coating with bovine serum albumin (BSA). They found that formalinized cells alone reacted strongly with some anti-BSA sera. They traced this to a sensitization of some erythrocytes by sheep serum albumin (SSA) and the presence of cross-reacting (anti-SSA) antibody in some anti-BSA sera. Sensitivity to anti-BSA could be increased, however, by tanning and adsorption of BSA. The tanned, formalinized red cell was used by ALBERGHINA and MAZZETTI (1969) as a carrier for the ribosomal proteins of *Neurospora crassa*. These sensitized cells were readily agglutinated by homologous antisera at fairly high titers. Red cells coated with ribosomal proteins from potato or yeast were not agglutinated by the antiserum against *N. crassa*. POLYAKOV et al. (1969) coated the tanned, formalinized red cell with *Brucella* antigens. They extensively investigated methods for preparing these hemagglutination indicators. DODIN et al. (1969) found the C fraction from the plague bacillus was its only protein antigen fixed on tanned, formalinized red cells and reacting with high titer antiplague sera. ZIMMERMAN et al. (1968) put the streptococcal M-protein on tanned red cells and obtained a sensitive indicator for type-specific antibody present after natural or induced streptococcal infections. This microtiter hemagglutination system eliminated the cumbersome procedures that had previously been needed to prevent errors due to cross-reactions.

A novel method of placing haptenic groups on an erythrocyte, without damaging the cells or causing nonspecific reactions with normal sera, was described by SORG et al. (1969). The hapten was placed on stearoaminoethyl poly-(vinyl alcohol) and this was then readily adsorbed to the red cell.

COOMBS et al. (1967) used an immunochemical linkage to attach antigens to red cells. The antigen was linked to a rabbit antibody against the red cell in such a manner that the antibody no longer agglutinated the red cells but still attached to them. This indicator was particularly useful in detecting incomplete antibodies, and could be designed to reveal their immunoglobulin type. Details of the coupling techniques were described by HUNTER and COOMBS (1969). Photooxidation coupled antibody fairly well to BSA, β-lactoglobulin (BL), castor bean protein, or α-lactalbumin, but only poorly coupled it to human serum albumin (HSA), egg albumin or pollen proteins. BDB or 2,4-diisocyano-toluene gave better coupling to BSA, BL, HSA or pollen extract, but this coupled antibody still agglutinated the erythrocytes in the absence of the specific antibody to the conjugated proteins. Carbodiimide coupling was notorious in producing aggregated material of poor utility.

Quantitative determination of antigens chemically bound to the surface of red blood cells was described by BORSOS et al. (1968), using the complement C'1a fixation and transfer test. They found that IgM generated complement-fixing sites on the red cell faster than did IgG, and generated twice as many sites.

Carriers other than the red blood cell also give useful indicators. Phage inactivation by reaction with an antibody against a hapten or antigen coupled to the phage gives a readily detected end point. This technique has been considerably exploited at the Weizmann Institute. HAIMOVICH et al. (1967) detected anti-penicilloyl activity by its inactivation of penicilloylated bacteriophage. Prior to this, HAIMOVICH and SELA (1966) had described similar results with poly-DL-alanylated bacteriophage. This and the paper of MAKELA (1966) are probably the first reports of the use of antigen (hapten) on phage, although there was considerable previous work with phage-antiphage systems. HAIMOVICH (1969) further described use of the penicilloylated bacteriophage T4 for detection of the reaginic anti-penicilloyl antibodies in allergic individuals. HAIMOVICH et al. (1969b) and HAIMOVICH and SELA (1969a), in their work on immunosorbents for isolating the antibodies, described the modified phages used in detecting these antibodies. HAIMOVICH and SELA (1969b) attached antibodies to phage and detected antigens, and in this paper also summarized much of their work with antigens attached to phage. HAIMOVICH et al. (1969a) recently delivered a paper on the inhibition of the antibody-inactivation of phage-antigen complexes.

The phage as an easily detected carrier was also used by others. BARBER and RITTENBERG (1969) used picrylated phage T4, and compared the kinetics of its inactivation by anti-picryl antibody to the kinetics of inactivation by antiphage. HORNICK and KARUSH (1969) conducted similar studies with 2,4-dinitrophenylated (DNP) phage ΦX174. KARUSH and HORNICK (1969) used this modified phage to detect antibody naturally bound to lymphoid cells. Cells with bound antibody fixed the DNP-phage and formed plaques when plated along with *E. coli*. However, control spleen cells often adsorbed considerable phage and improvement in the assay was still needed. FUCHS and FUCHS (1969) assayed the plant hormones, indoleacetic acid and gibberellic acid, by their inhibition of the inactivation by anti-hormones of a phage-hormone complex. OLOVNIKOV (1964c) used diazotized benzidine to prepare benzidine-protein complexes agglutinated by antibody to the complexed protein. OLOVNIKOV (1966, 1967) also described the use of tetrazotized p-diaminodiphenylamine to polycondense antigens or to couple them to red cells.

SLOBIN (1970) examined a hapten derivative of tobacco mosaic virus (TMV), prepared by reaction of TMV with methyl picolinimidate. This P-TMV behaved as a simple univalent hapten, inhibiting homologous precipitin reactions, but failing to precipitate with the antibody. The indicator described by R. B. FRITZ et al. (1968) was hapten diazotized to an unreactive carrier protein. Antibody to the hapten could be detected by its precipitation of this hapten-carrier complex. The radioimmune assay of antibodies to the C-terminal tetrapeptide amide (GT) of gastrin was accomplished by JAFFE et al. (1969), using GT coupled to a random copolymer of glutamate, alanine and tyrosine, and then labeled with ^{125}I. Thus one can devise a radioimmune assay of small molecules which cannot be iodinated by themselves. Antibody binding to gastrin and cholecystokinin-pancreozymin was also detected by this system. MIDGLEY et al. (1969) described a similar radioimmune assay of haptens, made possible by radio-iodination of the protein to which the hapten was conjugated. They used such a system for determination of estrogenic steroids.

The indicator used by KLINMAN and TAYLOR (1969) was antigen adsorbed on polystyrene latex. Antibody, in amounts as small as 0.1 ng, was detected after its reaction with this antigen-latex by adding ^{131}I-labeled anti-Ig and counting the γ-rays. Autoradiography was also applied with this indicator system.

Mathot and Rothen (1969a) described studies related to immunoelectroadsorption in which antigen anchored on a solid slide was used to determine rates of specific antibody adsorption. They examined systems in which the polysaccharide of type 3 pneumococcus or the human globulins were the anchored antigens.

E. Bound Antigens for Immunogenicity

The immunogenicity of substances can frequently be enhanced or altered by coupling them to larger molecules or particulate matter. This is necessary for haptenic substances and so much work has been done in this area that it is probably foolish to list only a few of these papers. Some of this work, however, was closely related to other subject matter of this review, often being extensions of other work on anchored biopolymers by the same authors.

Draper and Hirata (1968) compared the immunogenicity of soluble and particulate bovine serum albumin when injected into rabbits. Olovnikov and Gurvich (1966) coupled antigens to cellulose and induced greater and more specific antibody response in rabbits than was obtained by using soluble antigens and adjuvants. The British patent of Behringwerke A.-G. (1967) described a process for manufacture of artificial antigens. The p-nitrobenzyl derivative of antigen was reduced, diazotized, coupled to albumin, crosslinked with toluene diisocyanate, washed and lyophilized.

Arnon et al. (1967) used the linear copolymer of alanine, glutamic acid, and tyrosine as the body of an immunogen onto which lactosyl sphingosine groups were attached. This synthetic immunogen induced rabbits to produce antibody which precipitated the natural glycolipid called cytolipin H. A β-lactoside conjugate of poly-L-lysine was used as an immunogen by Gopalakrishnan and Karush (1969). Richards et al. (1969) used dinitrophenylated poly-(aminoacids) or protein. For their study of immunoindicators, Fuchs and Fuchs (1969) needed antibodies reactive with indoleacetic or gibberellic acids. They made these plant hormones immunogenic by conjugating them to keyhole limpet hemocyanin or bovine serum albumin. Bauminger et al. (1969a) made genistein immunogenic by conjugating it to a synthetic polypeptide. Midgley et al. (1969) made estrogenic steroids into immunogens by conjugating them to a protein. Spragg et al. (1968) made a copolymer of bradykinin and polylysine in order to induce antibodies to bradykinin in rabbits. Talamo et al. (1968) used toluene diisocyanate to couple bradykinin and ovalbumin, producing an immunogen for injection of animals to obtain antibody to bradykinin. The radioimmune assay described by Goodfriend et al. (1968) required antibody to angiotensin which they obtained by injecting rabbits with angiotensin coupled to albumin. A similar assay involving a tetrapeptide from gastrin (GT) required conjugation of the GT to protein via carbodiimides, yielding the immunogen which Jaffe et al. (1969) injected into their rabbits.

Fischer et al. (1969) examined several carriers in immunogens with bradykinin haptens. Antiserum induced by bradykinin-polylysine was homogeneous but the conjugate with ovalbumin elicited a mixture, as did that with human kininogen I. MAURER and PINCHUCK (1967) examined the antigenicity of poly-(amino acids). Linear copolymers of D-amino acids were not immunogenic and required a carrier such as methylated bovine serum albumin. Linear copolymers of L-amino acids were immunogenic without a carrier.

The group A polysaccharide of streptococci was modified and conjugated by I. GOLDSTEIN and CARAVANO (1968) via diazo linkages to edestin. This complex was stable for several months at 4°, and in rabbits induced antibody against both components. The anti-group A polysaccharide activity was specific, showing no cross-precipitation with extracts of streptococci from other groups.

Antibodies to porcine ACTH were obtained by REICHLIN et al. (1968) from rabbits immunized with ACTH and bovine serum albumin (BSA) cross-linked by glutaraldehyde. The immunogen was a potent inducer of antibodies, but had no steroidogenic activity. GROSS et al. (1968) rendered steroids immunogenic by a direct coupling through the phenolic A ring to diazotized p-aminobenzoic acid derivatives of BSA or hemocyanin. Slight cross-reactivity ensued among estradiol, estriol, and estrone. Testosterone and progesterone did not couple to the protein derivatives. ROTHENBERG et al. (1969) were able to produce antibody to folic acid by immunizing rabbits and guinea pigs with folic acid coupled to methylated bovine albumin. The antibody also reacted with Diopterin and Teropterin, but not with N^5-methyltetrahydrofolic acid nor with N^5-formyltetrahydrofolic acid. The antibody inhibited the reduction of folate to tetrahydrofolate and may have some utility as an inhibitor in studies of folate metabolism.

Extensive work has been done on the immunogenicity of polynucleotides. Some of this work was summarized at a 1967 symposium, and the papers from this symposium were published in a volume edited by PLESCIA and BRAUN (1968). Although RNA and its constituents are very poor immunogens, LACOUR (1968) developed agar diffusion techniques for their immunochemical detection, using antibody induced by injection of complexes such as ribosomes or RNA adsorbed or conjugated to proteins or polypeptides. Transfer RNA was adsorbed on methylated serum albumin (MeSA) and the complex induced anti-RNA antibodies. Uridine conjugated to polypeptides or to bovine serum albumin induced antibodies reacting with uridine and cross-reacting with RNA. Double-stranded poly A-poly U or poly I-poly C adsorbed on MeSA and injected into animals induced antibodies specific for the double stranded complexes. SELA and UNGAR-WARON (1968) prepared nucleoside immunogens by using a carbodiimide to couple the nucleosides to synthetic, nonantigenic, multichain polypeptides. The antibodies were specific for the entire nucleoside, in some cases, but for the constituent base in others. Some immunochemical specificity was also associated with the nature of the linkage between the base and the sugar components. They also described (UNGAR-WARON et al., 1967) not only the production of these antibodies, but immunosorbents for their isolation. BEISER et al. (1968) also summarized work in this immunochemical study of purine and pyrimidine compounds, particularly the reactions for coupling the purine, pyrimidines, ribosides, or ribotides to proteins in preparing the immunogens.

The antigenicity of synthetic polypeptides was examined by SIMONIAN et al. (1968). A polypeptide cross-linked by dinitrophenylene had unique antigenic sites dependent on the spatial organization of the molecule. Amide cross-links only slightly affected the immunogenicity and specificity of a polypeptide. All strains of rats respond to a D-amino acid polypeptide, but with interstrain differences in the magnitude of the response. Aggregation of L- or D-polypeptides in methylated serum albumin increased immunogenicity and reduced interstrain differences in the rat responses. HABEEB (1969) described changes in the antigenicity of bovine serum albumin (BSA) and ovalbumin (OA) after intramolecular crosslinking by formaldehyde or intermolecular crosslinking by glutaraldehyde. The aggregates after glutaraldehyde linking of BSA to OA induced rabbit antibodies with determinants for BSA, OA, and for glutaraldehyde-modified BSA or OA. ANDERER and SCHLUMBERGER (1969) enhanced the immunogenicity of proteins by cross linking them using diazonium derivatives of mixed polycondensates of aromatic amino carboxylic acids. Antibodies induced by the cross-linked protein also reacted with the protein monomer in the case of kallikrein inactivator, RNase, or human chorionic gonadotropin, and neutralized their biological activity. ANDERER and HOERNLE (1968) also cross-linked the kallikrein inactivator (KI) and used it as an immunogen. The antibodies reversed the action of KI. Effects of chemical alterations of KI on its antigenicity and biological activity were also explored.

ORLOV and GURVICH (1969) found γ-globulin in Sephadex was a much better immunogen than the globulin alone. Antibody levels in rabbits remained higher after primary injections of the Sephadex complex, and the response to a second injection was 15–20 times greater in these animals than in those receiving only γ-globulin.

The immunogenicity for rats of human serum albumin was enhanced when SCHENCK et al. (1969) conjugated it to the lipopolysaccharide of *E. coli*. Merely mixing the two substances was less effective. The conjugated material was as effective an immunogen in rats as the albumin in Freund's adjuvant or the albumin precipitated with alum, but Freund's adjuvant was the most effective in rabbits. The pyrogenicity of the lipopolysaccharide is undesirable, but this was greatly reduced by using succinylated lipopolysaccharide plus the polysaccharide component, without appreciable loss of the adjuvant effect.

Substances adsorbed on bentonite have an increased immunogenicity. GALLILY and GARVEY (1968a, 1968b) found hemocyanin poorly immunogenic, but a potent immunogen when adsorbed on bentonite. Macrophages more readily engulfed the particulate immunogen. They also noted (GALLILY and GARVEY, 1969) that the hemocyanin on bentonite caused a prolonged antibody response in the rats.

Voss et al. (1969) presented evidence that DNA could serve as a carrier for hapten-immunogens. Antibodies reactive with trinitrophenyl haptens were obtained after injection of DNA modified by 2,4,6-trinitrophenyl groups on its deoxyguanosine residues. ROELANTS and GOODMAN (1969) described RNA-antigen complexes, formed when RNA and antigen were mixed in the presence of a bivalent cation such as Mg^{2+}. SLOBIN (1970) concluded that virus-hapten conjugates induce anti-hapten antibodies only when the conformation of the carrier is altered by the attached hapten. The immunogenicity depended upon the extent of this change in conformation.

F. Related Studies of Antigen and Antibody

(Misc. Immunochemistry)

The immunochemistry of nucleic acids was recently reviewed by L. LEVINE and STOLLAR (1968). They particularly discussed the capacity of antibodies to detect unusual base pairs and structural changes in the nucleic acids. GARRO et al. (1969) described an immunochemical fractionation of DNA based on the use of anti-thymine and anti-cytosine.

Rabbit antibody to avian myoglobin was used by KAGEN (1969) to detect the areas in the chick embryo where myoglobin was most extensively produced. ISPOLATOVA (1968) found only those rat lymphoid cells that were actively producing antibody could react with erythrocytes to which the antigen had been conjugated. He concluded that the reaction was with antibody present on the surface of such lymphoid cells. PHILLIPS-QUAGLIATA et al. (1969a, 1969b) sought the mechanism by which phagocytes bind antigen-antibody complexes. They examined the binding by phagocytes of the anti-benzylpenicilloyl complexes with various types of haptens containing this determinant.

SPITLER et al. (1969) found migration of peritoneal exudate cells from capillaries was inhibited by their reaction with red cell-IgM complexes, but not by red cell-IgG. The hemocyanin complex with its antibody also inhibited the migration. MUELLER-ECKHARDT and LUESCHER (1969) described platelet aggregation, contraction, and release of adenine nucleotides when platelets contacted latex particles coated with γ-globulin. Latex coated with albumin or fibrinogen had no effect on the platelets. The γ-globulin configuration reacting with the platelets was the same as that for complement fixation.

WOJTYLA et al. (1969) titrated tetanus toxoid by adsorbing it on $AlPO_4$, reacting this adsorbed toxoid with fluorescein-labeled antitoxin, and determining the amount of bound fluorescence.

RUBALCAVA et al. (1969) used 1-anilinonaphthalene-8-sulfonate as a fluorescent probe on the erythrocyte membrane. They examined the effect of various agents on the membrane and thus on this probe, concluding that the probe interacts with hydrophobic-hydrophilic interphases containing neutral or hydrophilic groups. LALEZARI (1968) noted that Polybrene, a polymer with positive charges, enhanced the agglutination of erythrocytes, permitting their agglutination even by incomplete antibodies.

PASCOE (1969) examined the flocculation reaction of a polycation (polyethyleneamine) and a polyanion (protein) as a model of other systems, including the antigen-antibody reaction.

Miles and Hales (1968a) used insulin fixed to an insoluble matrix to bind anti-insulin, and then iodinated, protecting one or more active sites of the antibody during iodination. Day et al. (1967) protected active sites on antifibrin by iodinating the antibody while it was attached to fibrin. The dissertation of R. B. Fritz (1968) included the use of insoluble antigens in the study of the fibrinogen-antifibrinogen system. An unsigned leading article in Lancet (Anonymous, 1968) discussed the value of immunosorbents in the study of the reagenic activity of immunoglobulin IgE.

IV. Enzymes Bound to Antibodies

There have been many studies of cross-linked proteins, producing both soluble and insoluble complexes (Herzig et al., 1964; Habeeb and Hiramoto, 1968). Recently, a soluble ethylene-maleic anhydride derivative of trypsin has been prepared (Westman, 1969). Thus, bound protein complexes are not necessarily insoluble.

A very interesting soluble bound protein complex is that between an immunoglobulin and an enzyme. Nakane and Pierce (1966) covalently coupled γ-globulin to acid phosphatase and to peroxidase using a bifunctional reagent. These enzyme antibody conjugates possessed both enzymatic and immunologic activity, and were used for histochemical localization of tissue antigens. This was accomplished in an indirect manner by treating a tissue with the rabbit antibody to a particular tissue antigen, then with a peroxidase-labeled anti-rabbit-γ-globulin, and finally with a histochemical stain for peroxidase, which deposits a colored reaction product at the antigen site (Nakane and Pierce, 1966). The method is specific because of the antibody, and sensitive because of the amplifying effect of the enzyme. In addition, the antigen site can be located with an ordinary microscope or, if desired, with an electron microscope. A review covering materials of this nature has recently been published (Avrameas, 1970).

Specific examples employing this and similar techniques are numerous. For example, human serum albumin coupled to peroxidase was used to detect anti-albumin antibodies by immunoelectrophoresis (Avrameas and Uriel, 1966). Acid phosphatase and peroxidase linked to antibodies were used to immunohistochemically locate various tissue antigens. In this case, a direct procedure was used. The tissue was treated with an enzyme labeled antibody to the particular antigen of interest, and the material was histochemically stained for the enzyme (Nakane and Pierce, 1967).

Nakane et al. (1968) used the indirect method to locate the leucine-binding protein in *E. coli* cells. The fixed bacterial cells were treated with rabbit antibody to leucine binding protein, washed, and treated with peroxidase-labeled sheep anti-rabbit-γ-globulin. The cells were again washed, histochemically stained and examined in the electron microscope. The leucine binding protein was found only in the cell envelope of the bacteria.

Avrameas and Lespinats (1967) used cyanuric chloride to couple both human and bovine serum γ-globulin to peroxidase, glucose oxidase and acid phosphatase. These various preparations were compared for their ability to localize antibodies in spleen cells, and detect antibodies after immunoelectrophoresis. The spleens of rabbits hyperimmunized with human γ-globulin were treated with human γ-globulin-peroxidase conjugates in order to determine intracellular antibody sites (Bouteille and Avrameas, 1967; Avrameas and Bouteille, 1968). Similar studies were done with the spleens of rabbits hyperimmunized against other antigens (Leduc et al., 1969a). Casu and Avrameas (1969) prepared conjugates by coupling lactate dehydrogenase via glutaraldehyde to sheep anti-rabbit immunoglobulins. These were used for the detection of immunoglobulin in rabbit spleen cells. Using the same coupling agent, Avrameas (1969a) bound peroxidase, glucose oxidase, tyrosinase and alkaline phosphatase to human immunoglobulin G, human serum albumin, and to sheep and rabbit antibody.

Recently, Nakane (1970) used an indirect immunoenzymatic technique to locate six hormones in the anterior pituitary gland of the rat. The distribution of these hormones showed that growth hormone, adrenocorticotropic hormone, prolactin and thyrotropic hormone were found in separate cells, whereas follicle-stimulating hormone and luteinizing hormone were often found in the same cell. The distribution of the cells containing these hormones, as well as their ultrastructural characteristics were discussed. More recently, Kawarai and Nakane (1970) used the enzyme labeled antibody method to locate luteinizing hormone, growth hormone and prolactin on ultrathin sections of the rat anterior pituitary gland. Ultrathin sections were used in an effort to lessen the problems associated with diffusion of antibody and substrate. Sections were incubated in a steady flow of substrate, which reduced nonspecific staining and converted the technique into another useful tool for antigen localization. Similar examinations of ultrathin sections by previous workers were not as successful (Leduc et al., 1969a).

Schiff et al. (1970) used the indirect immunoenzymatic technique to study the distribution of chymotrypsinogen in pancreatic acinar cells. In addition, the enzyme labeled antibody method has been used in conjunction with classical staining techniques to correlate cell structure and function in pituitary sections (Swope et al., 1970). Vreeland (1970) has used the enzyme labeled antibody method to study the localization of a cell wall polysaccharide in algae. Guillien et al. (1970) used both immunofluorescent and immunoenzymatic techniques in an effort to study the specificity of cellular antibody production. Their results agreed with those of earlier workers in that only one type of antibody was produced by a single cell.

The indirect enzyme-labeled antibody method was extended by Nakane (1968) so that more than one tissue antigen could be located in a single tissue section. To illustrate this method, sections of rat pituitary glands were first treated with rabbit anti-human growth hormone, followed by peroxidase-labeled sheep anti-rabbit-γ-globulin and histochemically stained for peroxidase with a preparation which produced a yellow product. The section was washed and then treated with anti-human thyroid stimulating hormone, the peroxidase labeled anti-rabbit-γ-globulin, hydrogen peroxide and α-naphtholpyronin to give a pink product. The section was again washed, and then treated with anti-luteinizing hormone, peroxidase labeled anti-rabbit-γ-globulin, hydrogen peroxide and 4-chloro-1-naphthol to give a blue product. Each of the three antigens in the section could then be located by the color of the product specifically associated with that antigen. No more than one antigen was present in any one cell, and some cells contained none of these antigens.

Most applications of enzyme-labeled antibodies have involved the use of a coupling agent to bind the enzyme to the antibody. Recently, Avrameas (1969b) described other immunological complexes between the enzyme and antibody, which do not involve the use of coupling agents. One of these, the mixed antibody technique, uses a series of antigen-antibody reactions to detect an immunoglobulin in a specimen of a certain species by its reaction, in turn, with an anti-immunoglobulin from a second species, an anti-enzyme immunoglobulin of the first species, the enzyme, and finally a colorimetric substrate. A second technique, the hybrid method, uses an antibody of dual specificity. Part of the hybrid antibody reacts with immunoglobulin, the other part with enzyme. This hybrid antibody is prepared from a mixture of anti-immunoglobulin and anti-enzyme which has been partially digested, reduced, combined, and reoxidized. The enzyme site on the hybrid antibody in a section can be visualized by the addition of enzyme and colorimetric substrate for the enzyme. These methods would, of course, obviate any modification of the antibody or enzyme by the coupling agent, and they can be used in any situation where an enzyme labeled antibody would be used.

Enzyme labeled antibodies have also been used for localization of various virus antigens (Wicker and Avrameas, 1969; Leduc et al., 1969b; Kurstak et al., 1969). Thus, antigens of SV 40, adenovirus 12 and rat K viruses were detected. In one case, Wicker and Avrameas (1969) used two enzymes which produced different colored products and simultaneously detected two virus antigens in a single preparation.

V. Affinity Chromatography

Earlier in this review it was reported that insoluble inhibitors, and inhibitors bound to a modified ethylene maleic anhydride copolymer (Avrameas et al., 1969c; H. Fritz et al., 1969a) were used to isolate enzymes. These are examples of affinity chromatography, a purification method whereby an enzyme is isolated from a mixture by passing the mixture through a column to which a specific reactant for the desired enzyme had been covalently coupled. Ideally, the protein is retained or at least retarded by the column because of its interaction with the inhibitor. This allows all contaminating substances to be washed free from the column before the desired protein is eluted. Thus, affinity chromatography seems to possess great potential as a purification technique and deserves mention. There will be no attempt to completely cover the literature in this area. Pertinent examples will be selected to illustrate the method.

There were a few examples of enzyme purification in the earlier literature, using a support to which a molecule capable of complexing with the active site of the enzyme was attached (LERMAN, 1953; ARSENIS and McCORMICK, 1966). These columns, however, proved to be only moderately successful, probably because of the nature of the support material. For example, LERMAN (1953) obtained approximately a 61-fold purification of tyrosinase by using a cellulose-azo dye column. ARSENIS and McCORMICK (1964, 1966) used various flavin materials attached to cellulose to purify certain flavin enzymes.

In 1968, CUATRECASAS et al. (1968) introduced Sepharose as a matrix material in the very successful purification of staphylococcal nuclease, α-chymotrypsin, and carboxypeptidase A, and revived interest in affinity chromatography as a purification technique. This matrix material displayed highly desirable characteristics and gave superior results. Recent reviews (CUATRECASAS, 1970b; JERINA and CUATRECASAS, 1969) describe the extent to which Sepharose has been used for affinity chromatography. Interestingly, CUATRECASAS (1969b) has also been able to covalently bind insulin to Sepharose and showed that this derivative retained transport and antilipolytic activities. These findings suggest that interaction of the insulin molecule with surface membrane structures may account for the metabolic effects demonstrated by this molecule. This same derivative was used as an immunosorbent for anti-insulin purification, with very good results (CUATRECASAS, 1969a).

CUATRECASAS (1970a) described the preparation of a number of derivatives of agarose and polyacrylamide which allowed the attachment of a variety of ligands at various distances from the gel matrix, using covalent bonds which are easily cleaved from the gel backbone. These techniques lessened the steric hindrance between protein and support and provided a gentle means for removing very tightly bound proteins through removal of the ligand and, consequently, the attached protein. These new derivatives may help extend the applicability of this technique.

In most affinity chromatography applications a ligand bound to a matrix material is used to isolate a protein which has an affinity for the ligand. However, a recent paper (GIVOL et al., 1970) describes the use of a protein-Sepharose combination for the purification of peptides labeled with ligands which have affinity for the bound protein. Thus, a peptide from bovine pancreatic RNase labeled with uridine diphosphate was isolated using an RNase-Sepharose column, and a dinitrophenyl (DNP)-labeled peptide was isolated from a DNP labeled antibody by a Sepharose-anti DNP column.

Affinity chromatography was useful in many of the studies of staphylococcal nuclease. It was used for enzyme purifications, in studies on affinity labeling of the active site of the enzyme (CUATRECASAS et al., 1968; WILCHEK, 1970), for the purification of nuclease fragments (TANIUCHI and ANFINSEN, 1969), and the purification and study of the synthetic polypeptides corresponding or analogous to the P_2 fragment of the enzyme (ONTJES and ANFINSEN, 1969a, 1969b).

Many enzymes have been purified by affinity chromatography. For example, acetylcholinesterase has been purified using an [N-(ε-aminocaproyl)-p-aminophenyl] trimethylammonium bromide-Sepharose column (KALDERON et al., 1970). BERMAN and YOUNG (1971) describe the purification of this enzyme from both bovine erythrocyte membranes and the electric tissue of *Electrophorus electricus* using affinity chromatography. The specific activity of commercial trypsin was increased approximately 10% after chromatography on a ovomucoid-Sepharose column (FEINSTEIN, 1970). The isolation and purification of wheat proteases from wheat flour by a hemoglobin-Sepharose column was also carried out (CHUA and BUSHUK, 1969). The extraneous non-proteolytic materials were completely removed from the column by washing with buffer, and enzymes were released by 0.1 N acetic acid. An agarose p-aminophenylmercuric acetate column was used to purify a crystalline preparation of papain on the basis of its thiol content (SLUYTERMAN and WIJDENES, 1970). Two materials differing in activity and thiol content were obtained from such a column. One was approximately 20 times as active as the other. Chorismate mutase has been purified from a treated extract of *Claviceps paspali* strain Pb 156 using a Sepharose-L-trytophan column (SPROSSLER and LINGENS, 1970). L-tryptophan is actually an activator for this enzyme and not a competitive inhibitor. An effector bound to Sepharose has been used to purify the regulator enzyme, 3-deoxy-D-arabino-heptulosonate-7-phosphate synthetase (CHAN and TAKAHASHI, 1969). DEUTSCH and MERTZ (1970) obtained a 200 fold purification of plasminogen through the use of an L-lysine-Sepharose column. Bovine pancreatic ribonuclease A was purified by affinity chromatography, taking advantage of a specific inhibitor attached to Sepharose (WILCHEK and GORECKI, 1969). KATO and ANFINSEN (1969) used a Sepharose-RNase S protein column to help purify synthetic S peptides prepared by the Merrifield solid phase method and obtained a number of very similar products. A carboxy-methyl-cellulose-RNase column was used to isolate an RNase inhibitor from rat liver (GRIBNAU et al., 1970). Recently, ROBINSON et al. (1971) have used porous glass as the support for immobilization or affinity chromatography of enzymes.

VARDANIS (1969) used the affinity of concanavalin A for certain polysaccharides for quick isolation of glycogen bound enzymes from a liver homogenate. The homogenate yielded a number of such enzymes, and the technique gave a 25-fold increase in the yield of glycogen synthetase when compared to an established isolation procedure. The author suggests use of this method for the isolation of other polysaccharide-bound enzymes. DONNELLY and GOLDSTEIN (1970) used glutaraldehyde insolubilized concanavalin A in the isolation of polysaccharides and glycoproteins and have suggested the use of insolubilized concanavalin A in fractionating mixtures of these materials.

A rather novel use of affinity chromatography involved the separation of cells (TRUFFA-BACHI and WOFSY, 1970). Hapten covalently attached to polyacrylamide was used to separate cells capable of producing antihapten antibodies from those cells which do not possess the ability to bind and produce these antibodies. Antigens and antibodies have also been isolated using Sepharose columns. Blood group substance A was coupled to Sepharose and used to isolate anti-A phytohemagglutinin from *Vicia cracca* and anti-A from human serum (KRISTIANSEN et al., 1969). WOFSY and BURR (1969) used hapten-coupled Sepharose to purify a number of antihapten antibodies, and a Sepharose-staphylococcal nuclease column was used to isolate and fractionate antibodies to the enzyme (OMENN et al., 1970).

Affinity chromatography has also been extended to proteins which, although neither enzymes nor antibodies, can bind small molecules tightly. CUATRECASAS and WILCHEK (1968) used a biocytin-Sepharose column for a 4000-fold purification of avidin from egg white in a single step. Conversely, avidin coupled to Sepharose has been used to bind biotin or biotin containing peptides (BODANSZKY and BODANSZKY, 1970). BURSTEIN (1969) obtained testosterone binding globulin from plasma through the use of a Sepharose column to which an androstane derivative was coupled. In a similar manner, PENSKY and MARSHALL (1969) used a thyroxine-Sepharose column to separate thyroxine binding globulin from serum.

VI. Applications in Nucleic Acid Research

In recent years, various materials bound to supports have been used in nucleic acid research. For example, insoluble polynucleotides have been used to fractionate polynucleotide mixtures (GILHAM, 1964; GILHAM and ROBINSON, 1964), to isolate enzymes (LITMAN, 1968), as primers and templates for polymerases (JOVIN and KORNBERG, 1968), in assays (COZZARELLI et al., 1967), and in sequential analysis (WAGNER et al., 1969). Support materials have even been used in oligonucleotide synthesis (HAYATSU and KHORANA, 1966, 1967; CRAMER et al., 1966; MELBY and STROBACH, 1967; LETSINGER and MAHADEVAN, 1965, 1966; LETSINGER et al., 1967; BLACKBURN et al., 1967). Bound biopolymers appear to have great potential for applications in this field. Therefore, some of the applications for insoluble materials in nucleic acid research will be discussed.

GILHAM (1962) bound thymidine polynucleotides to cellulose via dicyclohexyl-carbodiimide. Through complementary base sequences, a column of this material selectively bound and separated deoxyadenosine hexanucleotide from thymidine hexanucleotide. The column also selectively bound 6% of the RNA from an *E. coli* amino acid transfer RNA mixture. Similarly, a number of synthetic polynucleotides were linked to phosphocellulose, placed in columns, and used to isolate complementary polynucleotides (ADLER and RICH, 1962). Thymidylate-cellulose displayed the ability to selectively bind adenosine, 5′-adenylate, and polyadenylate materials, presumably through hydrogen bonding of complementary bases (SANDER et al., 1966). Cellulose-linked polydeoxyribonucleotides and polyribonucleotides have also been prepared and used as selective chromatographic materials (ERHAN et al., 1965). BAUTZ and HALL

(1962) separated T4-specific RNA from *E. coli* RNA by chromatography on T4 DNA-cellulose columns. DNA-cellulose in a column was used to analyze complementary DNA strands found in phage infected cells (Merriam et al., 1970).

Gilham (1964) took advantage of the differences in stability of the base pairing complexes between the polynucleotides of a mixture and cellulose-bound polynucleotides to separate the components of the mixture. Thymidine polynucleotide-cellulose was used to isolate various polynucleotides from a digest of the RNA from bromegrass mosaic virus (Gilham and Robinson, 1964). Recently, Gilham (1968) presented new methods for binding nucleotides, polynucleotides and nucleic acids onto cellulose, by concentrating the reaction mixture within the cellulose fibers. Presumably these methods may extend the applications of bound biopolymers in this field.

Cozzarelli et al. (1967) used a bound polynucleotide in the assay of an enzyme that joined another polynucleotide strand to the one already linked to the cellulose.

A number of reports (Letsinger and Mahadevan, 1965, 1966; Letsinger et al., 1967; Blackburn et al., 1967) describe solid phase oligonucleotide synthesis in a manner analogous to the Merrifield synthesis of polypeptides (Merrifield, 1965). These papers describe the attachment of a monomer unit to an insoluble support, and the building of an oligonucleotide chain by the subsequent and sequential attachment of additional monomer units. In another variation, the polymer supporting the growing chain was soluble in the reaction medium (Hayatsu and Khorana, 1966, 1967; Cramer et al., 1966). This polymer, with the oligonucleotide chain attached, could be precipitated from solution and washed after the addition of each monomer unit. With the aid of the proper blocking groups, the above methods were used to synthesize a number of oligonucleotides. Polymer supports have also been used in the study of phosphorylation mechanisms (Blackburn et al., 1969). This work indicated that at least three phosphate residues must interact in a phosphorylation reaction.

Wagner et al. (1969) attached yeast sRNA to an insoluble support and subjected the polyribonucleotide to sequential analysis. Although the method failed after only a few nucleotides were determined, it suggests the possibility of a fast, efficient and possibly automated sequencing system.

There have also been a number of other specialized uses of bound biopolymers in the nucleic acid field. For example, Litman (1968) used a calf thymus DNA-cellulose preparation as an affinity column to purify *M. luteus* DNA polymerase from a partially purified enzyme preparation. In addition, a ribonculease has been bound to CM-cellulose and its properties investigated (Kuriyama and Egami, 1966). Also, an oligomer of deoxyribothymidylate bound to cellulose served as a biological model, displaying the capacity to act as a primer and template for *E. coli* DNA polymerase, a template for RNA polymerase, and a terminal transferase initiator (Jovin and Kornberg, 1968). Polynucleotide phosphorylase fixed to cellulose has been used as an insoluble enzyme to prepare large amounts of homopolynucleotides (HOFFMAN et al., 1970).

A few examples of the many ways in which bound biopolymers have been used for purification, characterization, detection, alteration, and even synthesis of biochemical materials in the fields of enzymology and immunology have been given. In this rapidly growing field, there will be many additional examples by the time this review is printed. Hopefully, the review may suggest even further application of these useful materials.

Acknowledgements. The authors are deeply indebted to Miss JACQUELINE M. MCCOMBS for her assistance in the preparation of this manuscript.

References

Anonymous: Reagin and IgE. Lancet **1968I**, 1131–1132.

ABRAHAM, G. E.: Solid-phase radioimmunoassay of 17β-estradiol. J. clin. Endocr. **29**, 866–870 (1969).

ADLER, A. J., RICH, A.: Cellulose columns containing polyribonucleotides and ribonucleic acids. J. Amer. chem. Soc. **84**, 3977–3979 (1962).

AKATSUKA, T., NELSON, O. E.: Starch granule-bound adenosine diphosphate glucose-starch glucosyltransferases of maize seeds. J. biol. Chem. **241**, 2280–2286 (1966).

ALBERGHINA, F. A. M., MAZZETTI, L.: Immunochemical studies of ribosomal proteins using passive hemagglutination techniques. Experientia (Basel) **25**, 1256–1257 (1969).

ALEXANDER, B.: Fibrinogen clottability and fibrin formation. Thrombos. Diathes. haemorrh. (Stuttg.), Suppl. **20**, 191–200 (1966).

ALEXANDER, B., ENGEL, A. M.: Water-insoluble thrombin. Meth. Enzymol. **19**, 978–985 (1970).

ALEXANDER, B., RIMON, A., KATCHALSKI, E.: Effects of water-insoluble trypsin derivatives on fibrinogen. Fed. Proc. **24**, 804–809 (1965).

ALEXANDER, B., RIMON, A., KATCHALSKI, E.: Action of water-insoluble trypsin derivatives on fibrinogen clottability. Thrombos. Diathes. haemorrh. (Stuttg.) **16**, 507–525 (1966).

AMOS, H.: Detection of antibodies in penicillamine sensitivity. Postgrad. med. J., Suppl. **1968** (Oct.), 27–30.

ANDERER, F. A., HOERNLE, S.: Chemical studies on kallikrein inactivator from bovine lung and parotid gland. Ann. N.Y. Acad. Sci. **146**, 381–385 (1968).

ANDERER, F. A., SCHLUMBERGER, H. D.: Antigenic properties of proteins cross-linked by multidiazonium compounds. Immunochemistry **6**, 1–10 (1969).

ARNOLD, W. N.: β-Fructofuranosidase from grape berries. III. Identity of the soluble and bound fractions. Biochim. biophys. Acta (Amst.) **128**, 196–198 (1966).

ARNON, R., SELA, M., RACHAMAN, E. S., SHAPIRO, D.: Cytolipin H-specific antibodies obtained with a synthetic polypeptide conjugate. Europ. J. Biochem. **2**, 79–83 (1967).

ARQUILLA, E. R.: Antigen-coupled microorganisms as reagents for the detection of antigens or antibodies. Fr. 1,545,886, 15 Nov. 1968; 16 pp.

ARSENIS, C., MCCORMICK, D. B.: Purification of liver flavokinase by column chromatography on flavine-cellulose compounds. J. biol. Chem. **239** (9), 3093–3097 (1964).

ARSENIS, C., MCCORMICK, D. B.: Purification of flavine mononucleotide-dependent enzymes by column chromatography on flavine phosphate cellulose compounds. J. biol. Chem. **241** (2), 330–334 (1966).

ASKENASE, P. W., LEONARD, E. J.: Solid phase radioimmunoassay of β1c-globulin. Fed. Proc. **28**, 501 (1969).

AVRAMEAS, S.: Coupling of enzymes to proteins with glutaraldehyde. Use of the conjugates for the detection of antigens and antibodies. Immunochemistry **6**, 43–52 (1969a).

AVRAMEAS, S.: Indirect immunoenzyme techniques for the intracellular detection of antigens. Immunochemistry **6**, 825–831 (1969b).

Avrameas, S.: Immunoenzyme techniques: enzymes as markers for the localization of antigens and antibodies. Int. Rev. Cytol. **27**, 349–385 (1970).

Avrameas, S., Bouteille, M.: Ultrastructural localization of antibody by antigen label with peroxidase. Exp. Cell Res. **53**, 166–176 (1968).

Avrameas, S., Brailovsky, C., Ternynck, T.: Isolation of viruses using specific immunoadsorbents. Virology **37**, 508–510 (1969a).

Avrameas, S., Lespinats, G.: Protein-bound enzymes, their use for the detection of antibodies and antigens. C. R. Acad. Sci. (Paris), Ser. D **265**, 1149–1152 (1967).

Avrameas, S., Taudou, B., Chuilon, S.: Glutaraldehyde, cyanuric chloride, and tetra-azotized o-dianisidine as coupling reagents in the passive hemagglutination test. Immunochemistry **6**, 67–76 (1969b).

Avrameas, S., Ternynck, T.: Utilization of the insoluble polymers of proteins for the isolation of antigens and antibodies. C. R. Acad. Sci., (Paris) Ser. C **262**, 1175–1177 (1966).

Avrameas, S., Ternynck, T.: Biologically active water-insoluble protein polymers. I. Their use for isolation of antigens and antibodies. J. biol. Chem. **242**, 1651–1659 (1967).

Avrameas, S., Ternynck, T.: Cross-linking of proteins with glutaraldehyde and its use for preparation of immunoadsorbents. Immunochemistry **6**, 53–66 (1969).

Avrameas, S., Ternynck, T., Guilbert, B.: Preparation of insoluble protein derivatives and their use in the isolation of antibodies, antigens, enzymes, and enzyme inhibitors. C. R. Acad. Sci. (Paris), Ser. D **268**, 227–229 (1969c).

Avrameas, S., Uriel, J.: Method for labeling antigens and antibodies with enzyme and their application in immunodiffusion C. R. Acad. Sci. (Paris), Ser. D **262**, 2543–2545 (1966).

Bachler, M. J., Strandberg, G. W., Smiley, K. L.: Starch conversion by immobilized glucoamylase. Biotech. Bioeng. **12**, 85–92 (1970).

Baker, P. J., Stashak, P. W., Prescott, B.: Use of erythrocytes sensitized with purified penumococcal polysaccharides for the assay of antibody and antibody-producing cells. Appl. Microbiol. **17**, 422–426 (1969).

Barber, P., Rittenberg, M. B.: Anti-trinitrophenyl (TNP) antibody detection by neutralization of TNP-coliphage T4. Immunochemistry **6**, 163–174 (1969).

Barfort, P., Arquilla, E. R., Vogelhut, P. O.: Resistance changes in lipid bilayers: immunological applications. Science **160**, 1119–1121 (1968).

Barker, S. A., Epton, R.: Water-insoluble enzymes. Process Biochem. **5** (8), 14–15, 19 (1970).

Barker, S. A., Somers, P. J., Epton, R.: Preparation and properties of alpha-amylase chemically coupled to microcrystalline cellulose. Carbohyd. Res. **8**, 491–497 (1968).

Barker, S. A., Somers, P. J., Epton, R.: Preparation and stability of exo-amylolytic enzymes chemically coupled to microcrystalline cellulose. Carbohyd. Res. **9**, 257–263 (1969).

Barker, S. A., Somers, P. J., Epton, R.: Recovery and re-use of water-insoluble amylase derivatives. Carbohyd. Res. **14**, 323–326 (1970a).

Barker, S. A., Somers, P. J., Epton, R., McLaren, J. V.: Cross-linked polyacrylamide derivatives (enzacryls) as water-insoluble carriers of amylolytic enzymes. Carbohyd. Res. **14**, 287–296 (1970b).

Bauman, E. K., Goodson, L. H., Guilbault, G. G., Kramer, D. N.: Preparation of immobilized cholinesterase for use in analytical chemistry. Analyt. Chem. **37**, 1378–1381 (1965).

Bauminger, S., Lindner, H. R., Perel, E., Arnon, R.: Antibodies to phytoestrogen-antigenicity of genistein coupled to a synthetic polypeptide. Israel J. med. Sci. **5**, 438 (1969a).

Bauminger, S., Lindner, H. R., Perel, E., Arnon, R.: Antibodies to a phyto-oestrogen; antigenicity of genistein coupled to a synthetic polypeptide. J. Endocr. **44**, 567–578 (1969b).

Bautz, E. K. F., Hall, B. D.: The isolation of T4-specific RNA on a DNA-cellulose column. Proc. nat. Acad. Sci. (Wash.) **48**, 400–408 (1962).

Beaumont, J. L., Delplanque, B.: Purification of rabbit anti-human β-lipoprotein antibodies from specific precipitates. Immunochemistry **6**, 489–494 (1969).

BEHRENS, M. M., INMAN, J. K., VANNIER, W. E.: Protein-cellulose derivatives for use as immunoadsorbents: preparation employing an active ester intermediate. Arch. Biochem. **119**, 411–419 (1967).

Behringwerke, A.G.: Manufacture of artificial antigens. Brit. 1,058,828 (Feb. 15, 1967).

BEISER, S. M., TANENBAUM, S. W., ERLANGER, B. F.: Purine- and pyrimidine-protein conjugates. Meth. Enzymol. **12** (Pt. B), 889–893 (1968).

BENNICH, H. H., JOHANSSON, S. G. O., WIDE, L. E.: Method for determining the presence of reagin immunoglobulins (reagin-Ig) directed against certain allergens. S. African 68 05,561, 5 Feb. 1969; 29 pp.

BERMAN, J. D., YOUNG, M.: Rapid and complete purification of acetylcholinesterases of electric eel and erythrocyte by affinity chromatography. Proc. nat. Acad. Sci. (Wash.) **68** (2), 395–398 (1971).

BERNFELD, P., WAN, J.: Antigens and enzymes made insoluble by entrapping them into lattices of synthetic polymers. Science **142**, 678–679 (1963).

BING, D. H., BURR, B., WOFSY, L.: Azo-hemocyanin conjugates as immunoadsorbents. Fed. Proc. **28**, 325 (1969).

BLACKBURN, G. M., BROWN, M. J., HARRIS, M. R.: Synthetic studies of nucleic acids on polymer supports. I. Oligodeoxyribonucleotide synthesis on an insoluble polymer support. J. chem. Soc. C **1967** (22), 2438–2442.

BLACKBURN, G. M., BROWN, M. J., HARRIS, M. R., SHIRÈ, D.: Synthetic studies of nucleic acids on polymer supports. II. Mechanisms of phosphorylation with carbodiimides and arene-sulfonyl chlorides. J. chem. Soc. C **1969** (4), 676–683.

BLUMENTHAL, R., CAPLAN, S. R., KEDEM, O.: The coupling of an enzymatic reaction to transmembrane flow of electric current in a synthetic "active transport" system. Biophys. J. **7**, 735–757 (1967).

BODANSZKY, A., BODANSZKY, M.: Sepharose-avidin column for the binding of biotin or biotin-containing peptides. Experientia (Basel) **26** (3), 327 (1970).

BORSOS, T., COLTEN, H. R., SPALTER, J. S., ROGENTHINE, N., RAPP, H. J.: The C'1a fixation and transfer test; its applicability to the detection and enumeration of antigens and antibodies at cell surfaces. J. Immunol. **101**, 392–398 (1968).

BOUTEILLE, M., AVRAMEAS, S.: Electron-microscopic study of antibody formation, by using peroxidase-labeled antigens. C. R. Acad. Sci. (Paris), Ser. D **265**, 2097–2099 (1967).

BOZICEVICH, J.: Detection of rubella virus (German measles) antibodies. U.S. 3,449,488, 10 June 1969; 4 pp.

BROUN, G., SELEGNY, E., AVRAMEAS, S., THOMAS, D.: Enzymatically active membranes. Some properties of cellophane membranes supporting cross-linked enzymes. Biochim. biophys. Acta (Amst.) **185**, 260–262 (1969).

BROUN, G., SELEGNY, E., TRAN MINH, C., THOMAS, D.: Facilitated transport of carbon dioxide across a membrane bearing carbonic anhydrase. FEBS (Fed. Eur. Biochem. Soc.) Letters **7** (3), 223–226 (1970).

BROWN, H. D., CHATTOPADHYAY, S. K., PATEL, A.: Sarcoplasmic reticulum ATPase on a solid support. Biochem. biophys. Res. Commun. **25**, 304–308 (1966a).

BROWN, H. D., CHATTOPADHYAY, S. K., PATEL, A.: Restructuring by linkage to insoluble matrices of enzymes solubilized from cell particles. J. Cell Biol. **31**, No 2, 17A (1966b).

BROWN, H. D., CHATTOPADHYAY, S. K., PATEL, A.: Characteristics of an ATPase in membrane particles, solubilized, and linked to a cellulose matrix. Enzymologia **32**, 205–212 (1967).

BROWN, H. D., PATEL, A. B., CHATTOPADHYAY, S. K.: Lattice entrapment of glycolytic enzymes. J. Chromatogr. **35**, 103–105 (1968).

BURSTEIN, S. H.: Removal of testosterone-binding globulin from plasma by affinity chromatography. Steroids **14**, 263–268 (1969).

BUTT, W. R., LYNCH, S. S.: Radioimmunoassay of gonadotropins, with special reference to follicle-stimulating hormone. Clin. chim. Acta **22**, 79–84 (1968).

CAMPBELL, D. H.: Isolation of protein antigens and antibodies. U.S. 2,957,808, Oct. 25, 1960.

CARPENTER, R. R., BARSALES, P. B.: Uptake by mononuclear phagocytes of protein-coated bentonite particles stabilized with a carbodiimide. J. Immunol. **98**, 844–853 (1967).

CARPENTER, R. R., BARSALES, P. B., GANCHAN, R. P.: Antigen-induced inhibition of cell migration in agar gel, plasma clot, and liquid media. RES, J. reticuloendoth. Soc. **5**, 472–483 (1968).

CARPENTER, R. R., REISBERG, M. A.: Carbodiimide-induced bentonite-antigen complexes: readily prepared immunoadsorbents. J. Immunol. **100**, 873–879 (1968).

CARREL, S., GERBER, H., BARANDUN, S.: Preparation of polyacrylamide gels which contain protein and their use as ligh capacity immunosorbents. Nature (Lond.) **221**, 385–386 (1969).

CASU, A., AVRAMEAS, S.: Conjugation of lactic dehydrogenase with proteins by use of glutaraldehyde. Detection of antigens in immunocompetent cells by means of conjugates. Ital. J. Biochem. **18**, 166–173 (1969).

CATT, K. J., HIRSCH, J., CASTELAN, D. J., NIALL, H. D., TREGEAR, G. W.: Radioimmunoassay of fibrinogen and its proteolysis products. Thrombos. Diathes. haemorrh. (Stuttg.) **20**, 1–6 (1968a).

CATT, K. J., NIALL, H. D., TREGEAR, G. W.: Solid phase radioimmunoassay of human growth hormone. Biochem. J. **100**, 31C–33C (1966).

CATT, K. J., NIALL, H. D., TREGEAR, G. W.: A disk method for solid-phase radioimmunoassay of proteins. Aust. J. exp. Biol. med. Sci. **45**, 703–706 (1967a).

CATT, K. J., NIALL, H. D., TREGEAR, G. W.: A solid phase disc radioimmunoassay for human growth hormone. J. Lab. clin. Med. **70**, 820–830 (1967b).

CATT, K. J., NIALL, H. D., TREGEAR, G. W.: Solid phase radioimmunoassay. Nature (Lond.) **213**, 825–827 (1967c).

CATT, K. J., NIALL, H. D., TREGEAR, G. W., BURGER, H. G.: Disk solid-phase radioimmunoassay of human luteinizing hormone. J. clin. Endocr. **28**, 121–126 (1968b).

CATT, K. J., TREGEAR, G. W.: Solid-phase radioimmunoassay in antibody-coated tubes. Science **158**, 1570–1572 (1967).

CEBRA, J. J.: Use of water-insoluble papain for structural studies of the products of antigen-antibody interaction. Symp. Fundamental Cancer Res. **16**, 220–237 (1962) (Pub. 1963).

CEBRA, J. J.: The effect of sodium dodecyl sulfate on intact and insoluble papain hydrolyzed immune globulin. J. Immunol. **92**, 977–990 (1964).

CEBRA, J. J., GIVOL, D., SILMAN, H. I., KATCHALSKI, E.: A two-stage cleavage of rabbit γ-globulin by a water-insoluble papain preparation followed by cysteine. J. biol. Chem. **236**, 1720–1725 (1961).

CHAN, W. W.-C., TAKAHASHI, M.: Affinity chromatography of a regulatory enzyme based on specific interaction with the effector. Biochem. biophys. Res. Commun. **37** (2), 272–277 (1969).

CHANG, T. M. S.: Semipermeable microcapsules. Science **146**, 524–525 (1964).

CHANG, T. M. S.: Semipermeable aqueous microcapsules (artificial cells)—with emphasis on experiments in an extracorporeal shunt system. Trans. Amer. Soc. artif. intern. Org. **12**, 13–19 (1966).

CHANG, T. M. S.: Microcapsules as artificial cells. Sci. J. **3** (No. 7), 62–67 (1967).

CHANG, T. M. S.: The *in vivo* effects of semipermeable microcapsules containing L-asparaginase on 6C3HED lymphosarcoma. Nature (Lond.) **229**, 117–118 (1971).

CHANG, T. M. S., JOHNSON, L. J., RANSOME, O. J.: Semipermeable aqueous microcapsules. IV. Non-thrombogenic microcapsules with heparin-complexed membranes. Canad. J. Physiol. Pharmacol. **45**, 705–715 (1967).

CHANG, T. M. S., MACINTOSH, F. C., MASON, S. G.: Semipermeable aqueous microcapsules. I. Preparation and properties. Canad. J. Physiol. Pharmacol. **44**, 115–128, 513 (1966).

CHANG, T. M. S., POZNANSKY, M. J.: Semipermeable microcapsules containing catalase for enzymic replacement in acatalasemic mice. Nature (Lond.) **218**, 243–245 (1968).

CHENG, W. C. Y., TALMAGE, D. W.: Reaction of immunoglobulin-coated bentonite and radioiodinated antigen as a basis for antibody detection and antigen binding. J. Immunol. **103**, 1385–1394 (1969).

CHIBATA, I., TOSA, T.: Preparation of insoluble enzymes. Tampakushitsu Kakusan Koso **11**, 23–29 (1966).

CHIBATA, I., TOSA, T.: Insoluble enzymes. Kobunshi **1967**, 16 (184), 823–832.

CHIDLOW, J. W., STEPHEN, J., SMITH, H.: Disulfide-linked antibodies as immunosorbents: adsorption of homologous antigen from solutions containing cytochrome c. Biochem. J. **104**, 4P–5P (1967).
CHIDLOW, J. W., STEPHEN, J., SMITH, H.: Desorption of antigens from disulphide-linked immunosorbents: Recovery of human serum albumin at near neutral pH. Nature (Lond.) **218**, 773–774 (1968).
CHUA, G. K., BUSHUK, W.: Purification of wheat proteases by affinity chromatography on hemoglobin-sepharose column. Biochem. biophys. Res. Commun. **37**, 545–550 (1969).
CHUBB, R. C.: Polyaminostyrene purified chicken antibodies. Nature (Lond.) **214**, 422–423 (1967).
CHUNG, S.-T., HAMANO, M., AIDA, K., UEMURA, T.: Studies on ATP deaminase. III. Water-insoluble ATP deaminase. Agr. Biol. Chem. (Tokyo) **32**, 1287–1291 (1968).
COOMBS, R. R. A., FEINSTEIN, A., JONAS, W. E., LACHMANN, P. J.: Use of the red cell-linked antigen-antiglobulin reaction to determine the immunoglobulin type of incomplete antibodies to soluble antigens. Symp. Ser. Immunobiol. Stand. **4**, 147–158 (1967).
COZZARELLI, N. R., MELECHEN, N. E., JOVIN, T. M., KORNBERG, A.: Polynucleotide cellulose as a substrate for a polynucleotide ligase induced by phage T4. Biochem. biophys. Res. Commun. **28** (4), 578–586 (1967).
CRAMER, F., HELBIG, R., HETTLER, H., SCHEIT, K. H., SELIGER, H.: Oligonucleotide synthesis with a soluble polymer as carrier. Angew. Chem. Intern. Ed. Engl. **5**, 601 (1966).
CROOK, E. M.: Enzymes attached to solid matrices. Biochem. J. **107**, 1P–2P (1968).
CROOK, E. M., BROCKLEHURST, K., WHARTON, C. W.: Cellulose-insolubilized enzymes. Meth. Enzymol. **19**, 963–978 (1970).
CUATRECASAS, P.: Insulin-sepharose. Immunoreactivity and use in the purification of antibody. Biochem. biophys. Res. Commun. **35**, 531–537 (1969a).
CUATRECASAS, P.: Interaction of insulin with the cell membrane. Primary action of insulin. Proc. nat. Acad. Sci. (Wash.) **63**, 450–457 (1969b).
CUATRECASAS, P.: Protein purification by affinity chromatography. Derivatizations of agarose and polyacrylamide beads. J. biol. Chem. **245** (12), 3059–3065 (1970a).
CUATRECASAS, P.: Agarose derivatives and purification by affinity chromatography. Nature (Lond.) **228**, 1327 (1970b).
CUATRECASAS, P., WILCHEK, M.: Single-step purification of avidin from egg white by affinity chromatography on biocytin-sepharose columns. Biochem. biophys. Res. Commun. **33**, 235–239 (1968).
CUATRECASAS, P., WILCHEK, M., ANFINSEN, C. B.: Selective enzyme purification by affinity chromatography. Proc. nat. Acad. Sci. (Wash.) **61**, 636–642 (1968).
CUATRECASAS, P., WILCHEK, M., ANFINSEN, C. B.: Affinity labeling of active site of staphylococcal nuclease. Reactions with bromoacetylated substrate analogues. J. biol. Chem. **244**, 4316–4329 (1969).
DANDLIKER, W. B., DE SAUSSURE, V. A., LEVANDOSKI, N.: Antibody purification at neutral pH utilizing immunospecific adsorbents. Immunochemistry **5**, 357–365 (1968).
DAVIS, R. V., BLANKEN, R. M., BEAGLE, R. J.: Diazotized m-aminobenzyloxymethyl-cellulose as the insoluble matrix for an immunoadsorbent used in purification of antigens and antibodies. Biochemistry **8**, 2706–2711 (1969).
DAVIS, W. C., DOUGLAS, S. D., PETZ, L. D., FUDENBERG, H. H.: Ferritin-antibody localization of erythrocyte antigenic sites in immunohemolytic anemias. J. Immunol. **101**, 621–627 (1968).
DAY, E. D., LASSITER, S., FRITZ, R. B.: Radioiodination of antibodies adsorbed to insoluble antigens. J. Immunol. **98**, 67–71 (1967).
DEUTSCH, D. G., MERTZ, E. P.: Plasminogen: purification from human plasma by affinity chromatography. Science **170**, 1095–1096 (1970).
DIETRICH, F. M., FRISCHKNECHT, H.: Analysis of antibodies against synthetic angiotensin II by the passive-hemagglutination technique. Nature (Lond.) **217**, 264–265 (1968).
DODIN, A., RICHAUD, J., BRYGOO, E. R., RAJENISON, S.: Conditioned hemagglutination reaction in plague bacillus in various hosts. Relation between precipitating and hemagglutinating antibodies. II. Protein antigens. Ann. Inst. Pasteur **117**, 369–377 (1969).

Doney, K., Hyslop, J., Johnson, A. G.: Purification of human renin and antirenin. Fed. Proc. **27**, 732 (1968).

Donnelly, E. H., Goldstein, I. J.: Glutaraldehyde-insolubilized concanavalin A: an adsorbent for the specific isolation of polysaccharides and glycoproteins. Biochem. J. **118** (4), 679–680 (1970).

Doscher, M. S., Richards, F. M.: The activity of an enzyme in the crystalline state; ribonuclease S. J. biol. Chem. **238**, 2399–2406 (1963).

Draper, L. R., Hirata, A. A.: Antibody responses in rabbits to soluble and particulate forms of bovine serum albumin. Immunology **15**, 23–30 (1968).

Drizlikh, G. I., Gurvich, A. E.: Use of antibodies bound to an antigen coated on insoluble particles for specific binding of various proteins. Biokhimiya **29**, 1054–1062 (1964).

Dubois, M. P.: Sensitization of erythrocytes to a protein antigen (sheep prolactin) by chromium chloride ($CrCl_3$). Constants relating to the coupling reaction studied by passive immunohemolysis. C. R. Acad. Sci. (Paris), Ser. D **268**, 2006–2009 (1969a).

Dubois, M. P.: Inhibition of passive immunohemolysis and its equation, studied on the prolactin-antiprolactin antibody couple. Application to the titration of antigen. C. R. Acad. Sci. (Paris), Ser. D **268**, 2154–2157 (1969b).

Dubois, M. P., Renoux, M., Renoux, G.: Immunohemolysis after chromium chloride sensitization. Application to a protein hormone and a bacterial glycoprotein. Ann. Inst. Pasteur **115**, 979 (1968).

Emmart, E. W., Bates, R. W.: The use of immunochemical binding and Sephadex filtration in procedures toward the purification of ovine and piscine "prolactin". Gen. comp. Endocr. **11**, 580–594 (1968).

Engel, A., Alexander, B.: Zymogen activation by insoluble trypsin and thrombin. Fed. Proc. **24**, 512 (1965).

Engel, A., Alexander, B.: Activation of chymotrypsinogen-A by thrombin preparations. Biochemistry **5**, 3590–3598 (1966).

Engel, A., Alexander, B., Pechet, L.: Activation of trypsinogen and plasminogen by thrombin preparations. Biochemistry **5**, 1543–1551 (1966).

Erhan, S., Northrup, L. G., Leach, F. R.: A method potentially useful for establishing base sequences in codewords. Proc. nat. Acad. Sci. (Wash.) **53** (3), 646–652 (1965).

Estermann, E. F., Conn, E. E., McLaren, A. D.: Influence of pH on particulate and soluble aconitase and glutamic dehydrogenase of *Lupinus albus*. Arch. Biochem. **85**, 103–108 (1959).

Evans, W. H., Mage, M. G., Peterson, E. A.: Immunoadsorption of cells to an antibody coated polyurethane foam. J. Immunol. **102**, 899–907 (1969).

Evreinova, T. N., Bailey, A.: Enzyme activity in coacervate systems. Polyphenoloxidase and coacervates. Dokl. Akad. Nauk SSSR**179**, 723–725 (1968).

Feinstein, G.: Purification of trypsin by affinity chromatography on ovomucoid-Sepharose resin. FEBS (Fed. Eur. Biochem. Soc.) Letters **7** (4), 353–355 (1970).

Fischer, J., Spragg, J., Talamo, R. C., Pierce, J. V., Suzuki, K., Austen, K. F., Haber, E.: Structural requirements for binding of bradykinin to antibody. III. Effect of carrier on antibody specificity. Biochemistry **8**, 3750–3757 (1969).

Freedman, M. H., Grossberg, A. L., Pressman, D.: Evidence for ammonium and guanidinium groups in the combining sites of anti-p-azobenzenearsonate antibodies: separation of two different populations of antibody molecules. J. biol. Chem. **243** (23), 6186–6195 (1968).

Fritz, H., Brey, B., Schmal, A., Werle, E.: Use of water-insoluble derivatives of trypsin-kallikrein inhibitor for isolation of kallikrein and plasmin. Hoppe-Seylers Z. physiol. Chem. **350**, 617–625 (1969a).

Fritz, H., Gebhardt, M., Fink, E., Schramm, W., Werle, E.: Use of water-insoluble enzyme resins with polyanionic and polyamphoteric resin matrixes for isolation of protease inhibitors. Hoppe-Seylers Z. physiol. Chem. **350**, 129–138 (1969b).

Fritz, H., Gebhardt, M., Meister, R., Illchmann, K., Hochstrasser, K.: Isolation of protease inhibitors with water-insoluble trypsin-cellulose resins. Hoppe-Seylers Z. physiol. Chem. **351**, 571–574 (1970).

Fritz, H., Hochstrasser, K., Werle, E., Brey, B., Gebhardt, M.: Identification and preparative separation of proteinase inhibitors and proteinases by means of water-insoluble enzvme and inhibitor resins. Fresenius' Z. anal. Chem. **243**, 452–463 (1968).

FRITZ, H., SCHULT, H., HUTZEL, M., WIEDEMANN, M., WERLE, E.: Protease inhibitors. IV. Isolation of protease inhibitors with water-insoluble enzyme-resin complexes. Hoppe-Seylers Z. physiol. Chem. **348**, 308–312 (1967).

FRITZ, R. B.: The combined use of soluble precipitating and related insoluble adsorbing antigens to measure the effects of iodination on antibody. The fibrinogen-antifibrinogen system. Diss. Abstr. B. **28**, 3617 (1968).

FRITZ, R. B., LASSITER, S., DAY, E. D.: Effect of iodination on the precipitin reaction. Anti-p-azophenyltrimethylammonium (anti-Ap) hapten system. Immunochemistry **5**, 557–565 (1968).

FRYDMAN, R. B., CARDINI, C. E.: Studies on the biosynthesis of starch. II. Some properties of the adenosine diphosphate glucose: starch glucosyltransferase bound to the starch granule. J. biol. Chem. **242**, 312–317 (1967).

FUCHS, S., FUCHS, Y.: Immunological assay for plant hormones using specific antibodies to indoleacetic acid and gibberellic acid. Biochim. biophys. Acta (Amst.) **192**, 528–530 (1969).

GALLAGHER, J. S., VOSS, E. W., JR.: Molecular weight of a purified chicken antibody. Immunochemistry **6**, 199–206 (1969).

GALLILY, R., GARVEY, J. S.: Triggering anti-hemocyanin antibody production in mice by injection of hemocyanin-coated bentonite particles. Fed. Proc. **27**, 563 (1968a).

GALLILY, R., GARVEY, J. S.: Primary stimulation of rats and mice with hemocyanin in solution and adsorbed on bentonite. J. Immunol. **101**, 924–929 (1968b).

GALLILY, R., GARVEY, J. S.: Prolonged antibody production in rats after single injection of KLH (keyhole limpet hemocyanin) on bentonite. Immunochemistry **6**, 766–767 (1969).

GALLOP, R. G. C., TOZER, B. T., STEPHEN, J., SMITH, H.: Separation of antigens by immunological specificity. Use of cellulose-linked antibodies as immunosorbents. Biochem. J. **101**, 711–716 (1966).

GALVEZ, G. E.: Specific adsorption of plant viruses by antibodies coupled to a solid matrix. Virology **28**, 171–187 (1966).

GARRO, A. J., ERLANGER, B. F., BEISER, S. M.: Immunochemical fractionation of DNA. Fed. Proc. **28**, 696 (1969).

GHETIE, V., SCHELL, H. D.: Agarose-based immunoadsorbents. Rev. Roum. Biochim. **5**, 65–67 (1968).

GILHAM, P. T.: Complex formation in oligonucleotides and its application to the separation of polynucleotides. J. Amer. chem. Soc. **84**, 1311–1312 (1962).

GILHAM, P. T.: The synthesis of polynucleotide-celluloses and their use in the fractionation of polynucleotides. J. Amer. chem. Soc. **86** (22), 4982–4985 (1964).

GILHAM, P. T.: The synthesis of celluloses containing covalently bound nucleotides, polynucleotides and nucleic acids. Biochemistry **7**, 2809–2813 (1968).

GILHAM, P. T., ROBINSON, W. E.: The use of polynucleotide-celluloses in sequence studies of nucleic acids. J. Amer. chem. Soc. **86** (22), 4985–4989 (1964).

GILL, T. J., III, BERNARD, C. F.: Quantitative micro-method for measuring antibody, utilizing the bromoacetylcellulose immunoadsorbent. Immunochemistry **6**, 567–571 (1969).

GIVAS, Sister JOAN, CENTENO, E. R., MANNING, M., SEHON, A. H.: Isolation of antibodies to the C-terminal heptapeptide of myoglobin with a synthetic peptide. Immunochemistry **5**, 314–318 (1968).

GIVOL, D., SELA, M.: Comparison of fragments of rabbit antibodies and normal γ-globulin by the peptide map technique. Biochemistry **3**, 451–458 (1964).

GIVOL, D., WEINSTEIN, Y., GORECKI, M., WILCHEK, M.: A general method for the isolation of labelled peptides from affinity-labelled proteins. Biochem. biophys. Res. Commun. **38**, 825–830 (1970).

GOLD, E. R.: Agglutination of red cells by $CrCl_3$ solutions: differential agglutinability of coated and uncoated cells from various donors. Vox Sang. (Basel) **14**, 147–150 (1968).

GOLD, E. R., FUDENBERG, H. H.: Chromic chloride: a coupling reagent for passive hemagglutination reactions. J. Immunol. **99**, 859–866 (1967).

GOL'DFEL'D, M. G., VOROB'EVA, E. S., POLTORAK, O. M.: Comparative study of catalase activity in adsorption layers of various types. Zh. Fiz. Khim. **40**, 2594–2596 (1966).

GOLDMAN, R., GOLDSTEIN, L., KATCHALSKI, E.: Water-insoluble enzyme derivatives and artificial enzyme membranes. In: Biochemical aspects of reactions on solid supports, p. 1–78, ed. by George R. Stark. New York· Academic Press 1971a.

Goldman, R., Kedem, O., Katchalski, E.: Papain-collodion membranes. II. Analysis of the kinetic behavior of enzymes immobilized in artificial membranes. Biochemistry **7**, 4518–4532 (1968).

Goldman, R., Kedem, O., Katchalski, E.: Kinetic behavior of alkaline phosphatase-collodion membranes. Biochemistry **10** (1), 165–172 (1971).

Goldman, R., Kedem, O., Silman, I. H., Caplan, S. R., Katchalski, E.: Papain-collodion membranes. I. Preparation and properties. Biochemistry **7**, 486–500 (1968b).

Goldman, R., Silman, H. I., Caplan, S. R., Kedem, O., Katchalski, E.: Papain membrane on a collodion matrix: Preparation and enzymatic behavior. Science **150**, 758–760 (1965).

Goldstein, G. B., Underdown, B. J., Heiner, D. C., Rose, B., Goodfriend, L.: Isolation of antibodies to gliadin and ragweed antigen E with the use of immunosorbents. Int. Arch. Allergy **36**, 573–581 (1969).

Goldstein, I., Caravano, R.: Preparation of an antiserum strictly specific against group A polysaccharide by means of a purified C polysaccharide complexed with edestin. Curr. Res. Group A Streptococcus, Proc. Symp. **1966**, 117–119 (Pub. 1968).

Goldstein, L.: Water insoluble derivatives of proteolytic enzymes. Meth. Enzymol. **19**, 935–962 (1970).

Goldstein, L., Katchalski, E.: Use of water-insoluble enzyme derivatives in biochemical analysis and separation. Fresenius' Z. anal. Chem. **243**, 375–396 (1968).

Goldstein, L., Levin, Y., Katchalski, E.: A water-insoluble polyanionic derivative of trypsin. II. Effect of polyelectrolyte carrier on the kinetic behavior of the bound trypsin. Biochemistry **3**, 1913–1919 (1964).

Goldstein, L., Pecht, M., Blumberg, S., Atlas, D., Levin, Y.: Water insoluble enzymes. Synthesis of a new carrier and its utilization for preparation of insoluble derivatives of papain, trypsin and subtilopeptidase A. Biochemistry **9**, 2322–2334 (1970).

Goodfriend, T. L., Ball, D. L., Farley, D. B.: Radioimmunoassay of angiotensin. J. Lab. clin. Med. **72**, 648–662 (1968).

Goodfriend, T. L., Ball, D. L., Updike, S.: Antibody in polyacrylamide gel. A solid phase reagent for radioimmunoassay. Immunochemistry **6**, 481–484 (1969).

Gopalakrishnan, P. V., Karush, F.: The synthesis and immunogenicity of a β-lactoside hapten conjugated to poly-L-lysine. Fed. Proc. **28**, 325 (1969).

Gottlieb, C. W., Lau, K.-S., Herbert, V.: Immunologic studies using antigen-coated charcoal: application to rheumatoid factor. Arthr. and Rheum. **10**, 199–203 (1967).

Graf, M. W., Uhr, J. W.: A method for the removal of specific antibody from immunized rabbits. Proc. Soc. exp. Biol. (N.Y.) **131**, 1231–1234 (1969).

Grant, G. F., Simon, M.: Use of radioactive antibodies for characterizing antigens and application to study of flagella synthesis. J. Bact. **95**, 81–86 (1968).

Greaves, M. F., Torrigiani, G., Roitt, I. M.: Blocking of the lymphocyte receptor site for cell mediated hypersensitivity and transplantation reactions by antilight chain sera. Nature (Lond.) **222**, 885–886 (1969).

Green, D. E., Jarnefelt, J.: Enzymes and biological organization. Perspect. Biol. Med. **2**, 163–184 (1959).

Green, M. L., Crutchfield, G.: Preparation of water-insoluble derivatives of rennin and chymotrypsin and their use in the hydrolysis of casein and the clotting of milk. Biochem. J. **115**, 183–190 (1969).

Gribnau, A. G. M., Schoenmakers, J. G. G., Van Kraaikamp, M., Bloemendal, H.: High purification of the RNase inhibitor from rat liver by affinity chromatography. Biochem. biophys. Res. Commun. **38** (6), 1064–1068 (1970).

Gross, S. J., Campbell, D. H., Weetall, H. H.: Production of antisera to steroids coupled to proteins directly through phenolic A ring. Immunochemistry **5**, 55–65 (1968).

Guilbault, G. G.: Enzymes in analytical chemistry. Analyt. Chem. **38**, 527R–536R (1966).

Guilbault, G. G.: Use of enzymes in analytical chemistry. Analyt. Chem. **40**, 459R–471R (1968).

Guilbault, G. G., Hrabankova, E.: An electrode for determination of amino acids. Analyt. Chem. **42** (14), 1779–1783 (1970a).

GUILBAULT, G. G., HRABANKOVA, E.: L-Amino acid electrode. Analyt. Letters **3** (2), 53–57 (1970b).
GUILBAULT, G. G., KRAMER, D. N.: Fluorimetric system employing immobilized cholinesterase for assaying anticholinesterase compounds. Analyt. Chem. **37**, 1675–1680 (1965).
GUILBAULT, G. G., MONTALVO, J. G., JR.: A urea-specific enzyme electrode. J. Amer. chem. Soc. **91**, 2164–2165 (1969).
GUILBAULT, G. G., MONTALVO, J. G., JR.: An enzyme electrode for the substrate urea. J. Amer. chem. Soc. **92** (8), 2533–2538 (1970).
GUILLIEN, P., AVRAMEAS, S., BURTIN, P.: Specificity of antibodies in single cells after immunization with antigens bearing several antigenic determinants. Study with a new paired staining technique. Immunology **18** (4), 483–491 (1970).
GURVICH, A. E.: Extraction of pure antibodies and determination of the absolute quantity of antibodies in serum by means of antigens bound to cellulose. Sovrem. Metody v Biokhim. (Moscow: Med.) Sb. **1964**, 73–87.
GURVICH, A. E., ISPOLATOVSKAYA, M. V., MYASOEDOVA, K. N.: Isolation and determination of diphtheria antibodies by means of antigens fixed on cellulose. Vop. med. Khim. **7**, 55–61 (1961).
GURVICH, A. E., KAPNER, R. B.: Quantitative determination of antibodies with the aid of paper precipitation reaction. Lab. Delo **4**, No 2, 23–26 (1958).
GURVICH, A. E., KUZOVLEVA, O. B., ORLOV, G. E.: Preparation of immunosorbents as a complex of protein antigens and Sephadex particles. Lab. Delo **1967**, 732–734.
GUTMAN, M., RIMON, A.: The activation of plasminogen. II. On the nature of the proactivator. Canad. J Biochem. **42**, 1339–1345 (1964).
GYENES, L., SEHON, A. H.: Tolylene 2,4-diisocyanate as a coupling agent in the passive hemagglutination reaction. Immunochemistry **1**, 43–48 (1964).
HABEEB, A. F. S. A.: Antigenicity of formaldehyde- and glutaraldehyde-treated bovine serum albumin and ovalbumin-bovine serum albumin conjugate. J. Immunol. **102**, 457–465 (1969).
HABEEB, A. F. S. A., HIRAMOTO, R.: Reaction of proteins with glutaraldehyde. Arch. Biochem. **126**, 16–26 (1968).
HABER, J. A., ROSE, N. R.: Comparative studies of indirect hemagglutination. Int. Arch. Allergy **34**, 303–311 (1968).
HABERMANN, E., RAKER, K. O.: A radioimmunoassay using cellulose-bound antibodies and its application to staphylococcal alpha-toxin. Naunyn-Schmiedebergs Arch. Pharmak. exp. Path. **263**, 216–217 (1969).
HABERMANN, E., RAKER, K. O., ZEUNER, G.: Solid-phase radioimmunassay and blood level of staphylococcal α-toxin. Naunyn-Schmiedebergs Arch. Pharmak. exp. Path. **262**, 165–182 (1969).
HAFERKAMP, O., SCHAFER, H., WESSEL, W., HSU, K. C.: Studies of the antigen-antibody reaction by electron microscopy. I. On the transfer of agglutinating antibodies from sensitized sheep erythrocytes to freshly added (unsensitized) red blood cells. Int. Arch. Allergy **36**, 298–316 (1969).
HAIMOVICH, J.: Use of an immunoadsorbent for the isolation of anti-immunoglobulin antibodies. Biochim. biophys. Acta (Amst.) **147**, 394–395 (1967).
HAIMOVICH, J.: Detection of anti-penicilloyl antibodies by inactivation of penicilloylated T4 bacteriophage. Int. Arch. Allergy **36**, 140–142 (1969).
HAIMOVICH, J., NOVIK, N., SELA, M.: Inhibition of inactivation of modified phage. Israel J. med. Sci. **5**, 438 (1969a).
HAIMOVICH, J., SCHECHTER, I., SELA, M.: Combining sites of IgG and IgM antibodies of poly-D-alanyl specificity. Europ. J. Biochem. **7**, 537–543 (1969b).
HAIMOVICH, J., SELA, M.: Inactivation of poly-DL-alanyl bacteriophage T4 with antisera specific toward poly-DL-alanine. J. Immunol. **97**, 338–343 (1966).
HAIMOVICH, J., SELA, M.: Inactivation of bacteriophage T4, of poly-D-alanyl bacteriophage, and of penicilloyl bacteriophage by immunospecifically isolated IgM and IgG antibodies. J. Immunol. **103**, 45–55 (1969a).
HAIMOVICH, J., SELA, M.: Protein-bacteriophage conjugates. Application in detection of antibodies and antigens. Science **164**, 1279–1280 (1969b).
HAIMOVICH, J., SELA, M., DEWDNEY, J. M., BATCHELOR, F. R.: Antipenicilloyl antibodies: detection with penicilloylated bacteriophage and isolation with a specific immunoadsorbent. Nature (Lond.) **214**, 1369–1370 (1967).

Hammerling, U., Aoki, T., De Harven, E., Boyse, E. A., Old, L. J.: Use of hybrid antibody with anti-γG and anti-ferritin specificities in locating cell surface antigens by electron microscopy. J. exp. Med. **128**, 1461–1473 (1968).

Hammerling, U., Aoki, T., Wood, H. A., Old, L. J., Boyse, E. A., De Harven, E.: New visual markers of antibody for electron microscopy. Nature (Lond.) **223**, 1158–1159 (1969).

Hammond, B. W., Kingsbury, D. T., Weiss, E.: Modification of meningococcal polysaccharide antigens for use in passive hemagglutination tests. J. Immunol. **101**, 808–809 (1968).

Hawkins, J. D.: A note on the use of formolized sheep red cells in the passive hemagglutination test for antisera to bovine serum albumin. Immunology **16**, 803–805 (1969).

Hayatsu, H., Khorana, H. G.: Deoxyribooligonucleotide synthesis on a polymer support. J. Amer. chem. Soc. **88**, 3182–3183 (1966).

Hayatsu, H., Khorana, H. G.: Studies on polynucleotides. LXXII. Deoxyribooligonucleotide synthesis on a polymer support. J. Amer. chem. Soc. **89**, 3880–3887 (1967).

Herzig, D. J., Rees, A. W., Day, R. A.: Bifunctional reagents and protein structure determination. The reaction of phenolic disulfonyl chlorides with lysozyme. Biopolymers **2**, 349–360 (1964).

Hicks, G. P., Updike, S. J.: The preparation and characterization of lyophilized polyacrylamide enzyme gels for chemical analysis. Analyt. Chem. **38**, 726–730 (1966).

Hirata, A. A.: Hemagglutination testing procedure with stabilized and potentiated erythrocytes and a method for their preparation. S. African 68, 06,993, 28 Mar. 1969; 23 pp.

Hirata, A. A., Brandriss, M. W.: Passive hemagglutination procedures for protein and polysaccharide antigens using erythrocytes stabilized by aldehydes. J. Immunol. **100**, 641–646 (1968).

Hirata, A. A., Sievert, H. W., Draper, L. R.: Adsorption of antibody on a metal surface. Fed. Proc. **28**, 308 (1969).

Hirata, A. A., Stall, W. T.: Reproducible passive hemagglutination procedure with high sensitivity using double aldehyde treated cells. Proc. Soc. exp. Biol. (N.Y). **130**, 343–345 (1969a).

Hirata, A. A., Stall, W. T.: Passive hemagglutination enhancement by periodate, low pH, and freeze-thaw treatment of aldehyde processed cells. Proc. Soc. exp. Biol. (N.Y.) **131**, 851–853 (1969b).

Hirata, A. A., Stashak, P. W.: Simplified procedure for passive hemagglutination tests with protein antigens. Nature (Lond.) **205**, 1233 (1965).

Hoff, G., Cohen, H.: Immunological determination of pregnant mare serum gonadotropin. Fed. Proc. **28**, 824 (1969).

Hoffman, C. H., Harris, E., Chodroff, S., Michelson, S., Rothrock, J. W., Peterson, E., Reuter, W.: Polynucleotide phosphorylase covalently bound to cellulose and its use in the preparation of homopolynucleotides. Biochem. biophys. Res. Commun. **41** (3), 710–714 (1970).

Hornby, W. E., Filippusson, H., McDonald, A.: Preparation of glucose oxidase chemically attached to polystyrene and its use in the automated analyses of glucose. FEBS (Fed. Eur. Biochem. Soc.) Letters **9** (1), 8–10 (1970).

Hornby, W. E., Lilly, M. D., Crook, E. M.: The preparation and properties of ficin chemically attached to carboxymethylcellulose. Biochem. J. **98**, 420–425 (1966).

Hornby, W. E., Lilly, M. D., Crook, E. M.: Some changes in the reactivity of enzymes resulting from their chemical attachment to water-insoluble derivatives of cellulose. Biochem. J. **107**, 669–674 (1968).

Hornick, C. L., Karush, F.: Interaction of hapten-coupled bacteriophage ΦX174 with antihapten antibody. Israel J. med. Sci. **5**, 163–170 (1969).

Hoyer, L. W., Vannier, W. E., Renfer, L.: Antibody elution from hapten-cellulose immunoadsorbents: the effects of hapten structure, pH, and salt concentration. Immunochemistry **5**, 277–292 (1968).

Hunter, A., Coombs, R. R. A.: Red cell-linked antigen-antiglobulin reaction. Int. Arch. Allergy **36**, 354–375 (1969).

Hussain, Q. Z., Newcomb, T. F.: Preparation of water-insoluble thrombin. Proc. Soc. exp. Biol. (N.Y.) **115**, 301–305 (1964).

INAI, S., TSUYUGUCHI, I.: Complement-dependent hemolysis of sheep erythrocytes coupled with nonantibody human immunoglobulin G. Biken J. (Biseibutsubyo Kenkyujo) **11**, 261–267 (1968).

ISPOLATOVA, A. V.: Adsorption of erythrocytes conjugated with soluble proteins. Byull. éksp. Biol. Med. **66**, 63–65 (1968).

ITO, K., WICHER, K., ARBESMAN, C. W.: Insoluble immunoadsorbents containing anti-IgE: removal of reaginic activity and subsequent elution. J. Immunol. **103**, 622–624 (1969).

JACKSON, G. D. F., ROWLEY, D., JENKINS, C. R.: Artificial antigens and immunity to mouse typhoid. I. Use of O-acetylated galactans in the purification of specific antibody against antigen 5. Immunology **15**, 789–798 (1968).

JAFFE, B. M., NEWTON, W. T., MCGUIGAN, J. E.: Antibodies to gastrin tetrapeptide: a new technique for radioimmunoassay. Fed. Proc. **28**, 501 (1969).

JAGENDORF, A. T., PATCHORNIK, A., SELA, M.: Use of antibody bound to modified cellulose as an immunospecific adsorbent of antigen. Biochim. biophys. Acta (Amst.) **78**, 516–527 (1963).

JANSEN, E. F., OLSON, A. C.: Properties and enzymatic activities of papain insolubilized with glutaraldehyde. Arch. Biochem. **129**, 221–227 (1969).

JAQUET, H., CEBRA, J. J.: Comparison of two precipitating derivatives of rabbit antibody; fragment I dimer and the product of pepsin digestion. Biochemistry **4**, 954–963 (1965).

JATON, J. C., KLINMAN, N. R., GIVOL, D., SELA, M.: Recovery of antibody activity on reoxidation of completely reduced polyalanyl heavy chain and its Fd fragment derived from anti-2,4-dinitrophenyl antibody. Biochemistry **7**, 4185–4195 (1968).

JERINA, D. M., CUATRECASAS, P.: Enzyme purification by affinity chromatography. Proc. IV Internatl. Congr. Pharmacol., Basel, Switz., July 1969.

JOHNSON, D. E., CIEGLER, A.: Substrate conversion by fungal spores entrapped in solid matrixes. Arch. Biochem. **130**, 384–388 (1969).

JONES, J. M., PERRY, S. V.: The biological activity of subfragment 1 prepared from heavy meromyosin. Biochem. J. **100** (1), 120–130 (1966).

JONIAU, M., ONKELINX, E., LONTIE, R.: Glutaraldehyde polymerized proteins as immunosorbents. Arch. int. Physiol. Biochim. **76**, 182–183 (1968).

JOVIN, T. M., KORNBERG, A.: Polynucleotide celluloses as solid state primers and templates for polymerases. J. biol. Chem. **243**, 250–259 (1968).

KAGEN, L. J.: Synthesis of myoglobin by chick embryos detected immunologically. Fed. Proc. **28**, 501 (1969).

KALDERON, N., SILMAN, I., BLUMBERG, S., DUDAI, Y.: A method for the purification of acetylcholinesterase by affinity chromatography. Biochim. biophys. Acta (Amst.) **207**, 560–562 (1970).

KAPLAN, M. E., KABAT, E. A.: Human antibodies. IV. Purification and properties of anti-A and anti-B obtained by absorption and elution from insoluble blood group substances. J. exp. Med. **123**, 1061–1081 (1966).

KARITZKY, D., KUNZER, W.: Preparation of a monospecific antiserum against human fibrinogen. Thrombos. Diathes. haemorrh. (Stuttg.) **18**, 532–536 (1967).

KARITZKY, D., KUNZER, W.: Preparation of an artificial fibrinogen deficit in human plasma. Klin. Wschr. **46**, 814–816 (1968).

KARLIN, A.: Association of acetylcholinesterase and membrane in subcellular fractions of the electric tissue of *Electrophorus*. J. Cell Biol. **25** (2, Part 2), 159–169 (1965).

KARUSH, F., HORNICK, C. L.: Plaque assay for cell-bound antihapten antibody. Cell. Recognition **1969**, 281–285.

KATCHALSKI, E.: A synthetic approach to the study of microenvironmental effects on enzyme action. "Structure-Function Relationships of Proteolytic Enzymes", ed. P. Desnuelle, H. Neurath, and M. Ottesen, p. 198–221. New York, N.Y.: Academic Press 1970.

KATO, I., ANFINSEN, C. B.: Purification of synthetic ribonuclease S-peptide derivatives by specific complex formation on columns of ribonuclease S-protein bound to agarose. J. biol. Chem. **244**, 5849–5855 (1969).

KATZ, Y., MAYER, A. M.: Changes in properties of catechol oxidase from chloroplasts following liberation from membranes. Israel J. Bot. **18**, 11–19 (1969).

KAWARAI, Y., NAKANE, P. K.: Localization of tissue antigens on the ultrathin sections with peroxidase-labeled antibody method. J. Histochem. Cytochem. **18** (3), 161–166 (1970).

KAY, G.: Insolublised enzymes. Process Biochem. **3**, No 8, 36–39 (1968).

KAY, G., LILLY, M. D., SHARP, A. K., WILSON, R. J. H.: Preparation and use of porous sheets with enzyme action. Nature (Lond.) **217**, 641–642 (1968).

KENT, L. H., SLADE, J. H. R.: Immunochemical polymers. Brit. 993,961 (June 2, 1965).

KING, C. A.: The adsorption of protein by tissue. Biochim. biophys. Acta (Amst.) **154**, 269–277 (1968).

KITAGAWA, M., YAGI, Y., PRESSMAN, D.: Heterogeneity of combining sites of antibodies as determined by specific immunoadsorbents. I. Rapid micromethod of fractionation of antibody. J. Immunol. **95**, 446–454 (1965a).

KITAGAWA, M., YAGI, Y., PRESSMAN, D.: Heterogeneity of combining sites of antibodies as determined by specific immunoadsorbents. II. Comparison of elution patterns obtained with anti-p-azobenzoate antibodies by different kinds of immunoadsorbent and eluting hapten. J. Immunol. **95**, 454–465 (1965b).

KLINMAN, N. R., TAYLOR, R. B.: Study of cells and serum during the immune response: response to dinitrophenyl in mice. Clin. exp. Immunol. **4**, 473–487 (1969).

KOMINZ, D. R., MITCHELL, E. R., NIHEI, T., KAY, C. M.: The papain digestion of skeletal myosin A. Biochemistry **4**, 2373–2382 (1965).

KOTOVA, T. S.: Preparation of an immunosorbent from cellulose. Lab. Delo **1966**, 93–96.

KOTOVA, T. S., DALIN, M. V.: Preparation of immunoadsorbents from amino celluloses. Zh. Mikrobiol. (Mosk.) **44**, 109–114 (1967).

KRISTIANSEN, T., PORATH, J.: Blood group substances. I. Purification of active material from hog gastric mucin by specific precipitation with *Vicia cracca* phytohemagglutinin. Biochim. biophys. Acta (Amst.) **158**, 351–357 (1968).

KRISTIANSEN, T., SUNDBERG, L., PORATH, J.: Studies on blood group substances. II. Coupling of blood group substance A to hydroxyl-containing matrices, including aminoethyl cellulose and agarose. Biochim. biophys. Acta (Amst.) **184**, 93–98 (1969).

KURIYAMA, Y., EGAMI, F.: Preparation and properties of water-insoluble ribonuclease T_1. Seikagaku **38**, 735–740 (1966).

KURSTAK, E., COTE, J. R., BELLONCIK, S.: Synthesis and localization of antigens of densonucleose virus using antibodies conjugated with peroxidase enzyme (new method of immunoperoxidase). C. R. Acad. Sci. (Paris), Ser. D **268**, 2309–2312 (1969).

KUZOVLEVA, O. B., GURVICH, A. E.: Synthesis and properties of new immunosorbents. Vop. med. Khim. **12**, 316–324 (1966).

KUZOVLEVA, O. B., GURVICH, A. E., ROGOVIN, Z. A., VIRNIK, A. D.: Synthesis of new immunosorbents based on the simple ether of cellulose and 4-(β-hydroxyethylsulfonyl)-2-aminoanisole. Vop. med. Khim. **12**, 106–107 (1966).

LACOUR, F.: Immunological detection of ribonucleic acids by agar diffusion. Meth. Enzymol. **12**, Pt. B, 903–907 (1968).

LALEZARI, P.: Detection of red blood cell antibodies. Transfusion (Philad.) **8**, 372–380 (1968).

LEDUC, E. H., SCOTT, G. B., AVRAMEAS, S.: Ultrastructural localization of intracellular immune globulins in plasma cells and lymphoblasts by enzyme-labelled antibodies. J. Histochem. Cytochem. **17**, 211–224 (1969a).

LEDUC, E. H., WICKER, R., AVRAMEAS, S., BERNHARD, W.: Ultrastructural localization of SV-40-T antigen with enzyme-labelled antibody. J. gen. Virol. **4**, 609–614 (1969b).

LERMAN, L. S.: A biochemically specific method for enzyme isolation. Proc. nat. Acad. Sci. (Wash.) **39**, 232–236 (1953).

LETSINGER, R. L., CARUTHERS, M. H., JERINA, D. M.: Reactions of nucleosides on polymer supports. Synthesis of thymidylylthimidylylthymidine. Biochemistry **6**, 1379–1388 (1967).

LETSINGER, R. L., MAHADEVAN, V.: Nucleotide chemistry. II. Oligonucleotide synthesis on a polymer support. J. Amer. chem. Soc. 87 (15), 3526–3527 (1965).

LETSINGER, R. L., MAHADEVAN, V.: Stepwise synthesis of oligodeoxyribonucleotides on an insoluble polymer support. J. Amer. chem. Soc. **88** (22), 5319–5324 (1966).

LEVIN, Y., PECHT, M., GOLDSTEIN, L., KATCHALSKI, E.: A water-insoluble polyanionic derivative of trypsin. I. Preparation and properties. Biochemistry **3**, 1905–1913 (1964).

LEVINE, B. B., LEVYTSKA, V.: Hapten-specific insoluble immunoabsorbents prepared from hide powder (collagen). J. Immunol. **102**, 647–655 (1969).

LEVINE, L., STOLLAR, B. D.: Nucleic acid immune systems. Progr. Allergy **12**, 161–191 (1968).

LILLY, M. D., HORNBY, W. E., CROOK, E. M.: The kinetics of carboxymethyl cellulose-Ficin in packed beds. Biochem. J. **100**, 718–723 (1966).

LILLY, M. D., KAY, G., SHARP, A. K., WILSON, R. J. H.: The operation of biochemical reactors using fixed enzymes. Biochem. J. **107**, 5P (1968).

LILLY, M. D., SHARP, A. K.: The kinetics of enzymes attached to water-insoluble polymers. Chem. Eng. (Lond.) No 215, CE12–CE18 (1968).

LINSCOTT, W. D., FAULK, W. P., PERUCCA, P. J.: Complement: chemical coupling of a functionally active component to erythrocytes in the absence of antibody. J. Immunol. **103**, 474–479 (1969).

LITMAN, R. M.: A deoxyribonucleic acid polymerase from *Micrococcus luteus* (*Micrococcus lysodeikticus*) isolated on deoxyribonucleic acid-cellulose. J. biol. Chem. **243**, 6222–6233 (1968).

LOEWY, S., GOLDSTEIN, L., COHEN, C., LUCK, S. M.: Proteolytic degradation of myosin and the meromyosins by a water-insoluble polyanionic derivative of trypsin. Properties of a helical subunit isolated from heavy meromyosin. J. molec. Biol. **23**, 287–304 (1967).

LOWEY, S., GOLDSTEIN, L., LUCK, S. M.: Isolation and characterization of a helical subunit from heavy meromyosin. Biochem. Z. **345**, 248–254 (1966).

MAGE, M. G., EVANS, W. H., PETERSON, E. A.: Enrichment of antibody plaque-forming cells by immunoadsorption to an antigen-coated polyurethane foam. J. Immunol. **102**, 908–910 (1969).

MAKELA, O.: Assay of anti-hapten antibody with the aid of hapten-coupled bacteriophage. Immunology **10**, 81–86 (1966).

MANECKE, G.: Serologically active protein resins and enzyme resins. Naturwissenschaften **51**, 25–34 (1964).

MANECKE, G.: Methods of attaching enzymes to various polymers and the effects on the properties of these enzymes. Biochem. J. **107**, 2P–3P (1968).

MAN'KO, N. I., VOLYNSKAYA, E. M., DARCHUK, G. F., BARANOVA, N. P.: Preparation of immunoadsorbents with the aid of isothiocyanopolystyrene. Vop. Immunol. **3**, 168–171 (1968).

MANN, D. L., GRANGER, H., FAHEY, J. L.: Use of insoluble antibody for quantitative determination of small amounts of immunoglobulin. J. Immunol. **102**, 618–624 (1969).

MATHEKA, H. D., MUSSGAY, M., Activated Sepharose as an adsorbent for antibodies of foot-and-mouth disease virus, and its use as an immunoadsorbent for type and subtype differentiation. Arch. ges. Virusforsch. **27**, 13–22 (1969).

MATHOT, C., ROTHEN, A.: Rates of specific adsorption of antibodies on layers of antigen anchored to a solid slide. 1. Polysaccharide of type 3 pneumococcus and homologous antibodies. 2. Human globulin and homologous antibodies. Biophys. J. **9**, A50 (1969a).

MATHOT, C., ROTHEN, A.: Adsorptions of purified serum proteins on metallized glass slides and their significance for the immunoelectroadsorption method. J. Colloid Interface Sci. **31**, 51–60 (1969b).

MATHOT, C., ROTHEN, A.: Immunoelectroadsorption method. Surface Sci. **16**, 428 (1969c).

MATTE, C., AUDRAN, R., MOULLEC, J., STEINBUCH, M.: Differential adsorption of rheumatoid factor and anti-Gm(a) activity. Rev. franç. Étud. clin. biol. **10**, 308–315 (1965).

MATTIOLI, C. A., YAGI, Y., PRESSMAN, D.: Production and properties of mouse antihapten antibodies. J. Immunol. **101**, 939–948 (1968).

MATUHASI, T., KOBAYASHI, T.: Immunological reaction. Brit. 1,155,365, 18 June 1969a; 5 pp.

MATUHASI, T., KOBAYASHI, T.: Method for detecting antigens. U.S. 3,457,344, 22 July 1969b; 3 pp.

MAURER, P. H., PINCHUCK, P.: Antigenicity of polypeptides (poly α-amino acids). XXIII. Immunogenicity of synthetic polymers of aminoacids; role of carrier and genetic background. Immunity, Cancer, Chemother., Symp. **1967**, 319–349.

MAZIA, D., HAYASHI, T.: The activity of pepsin-albumin films. Arch. Biochem. **43**, 424–442 (1952).

McDuffie, N. M., Sutherland, G. B.: Immunochemistry of coagulated antigen-antibody complex. I. General procedure and effects of coagulating conditions. J. Immunol. **90**, 438–443 (1963).

McLaren, A. D.: pH dependence of enzyme reaction on cells, particulates and in solution. Science **125**, 697 (1957).

McLaren, A. D.: Enzyme action in structurally restricted systems. Enzymologia **21**, 356–364 (1960).

McLaren, A. D., Babcock, K. L.: Characteristics of enzyme reactions at surfaces. "Subcellular Particles", Symposium, Woods Hole, Mass. 1958, p. 23–36 (Hayashi, ed.). New York: Ronald Press 1959.

McLaren, A. D., Packer, L.: Some aspects of enzyme reactions in heterogeneous systems. Advanc. Enzymol. **33**, 245–308 (1970).

Melby, L. R., Strobach, D. R.: Oligonucleotide synthesis on insoluble polymer supports. I. Stepwise synthesis of trithymidine diphosphate. J. Amer. chem. Soc. **89**, 450–453 (1967).

Merriam, E. V., Dumas, L. B., Sinsheimer, R. L.: A method for analysis of DNA by annealing on a DNA cellulose column. Analyt. Biochem. **36**, 389–395 (1970).

Merrifield, R. B.: Automated synthesis of peptides. Science **150**, 178–185 (1965).

Midgley, A. R., Jr., Niswender, G. D., Ram, J. S.: Hapten radioimmunoassay; general procedure for the estimation of steroidal and other haptenic substances. Steroids **13**, 731–737 (1969).

Mihaescu, S., Cilievici, D.: Isolation of pure antibodies by means of a fine PAB (p-aminobenzyl)-cellulose suspension. Rev. Roum. Biochim. **6**, 51–56 (1969).

Miles, L. E. M., Hales, C. N.: The preparation and properties of purified iodine-125-labeled antibodies to insulin. Biochem. J. **108**, 611–618 (1968a).

Miles, L. E. M., Hales, C. N.: Labeled antibodies and immunological assay systems. Nature (Lond.) **219**, 186–189 (1968b).

Mitz, M. A.: Insoluble active derivative of an enzyme as a model for study of cellular metabolism. Science **123**, 1076–1077 (1956).

Moreno, C., Kabat, E. A.: Human antibodies. VIII. Properties and association constants of human antibodies to blood group A substance purified with insoluble specific adsorbents and fractionally eluted with mono- and oligosaccharide. J. exp. Med. **129**, 871–896 (1969).

Mosbach, K.: Enzymes bound to artificial matrixes. Sci. Amer. **224** (3), 26–33 (1971).

Mosbach, K., Larsson, P. O.: Preparation and application of polymer-entrapped enzymes and microorganisms in microbial transformation processes with special reference to steroid 11-β-hydroxylation and Δ^1-dehydrogenation. Biotech. Bioeng. **12** (1), 19–27 (1970).

Mosbach, K., Mattiasson, B.: Matrix bound enzymes. II. Studies on a matrix bound two enzyme system. Acta chem. scand. **24**, 2093 (1970).

Mosbach, K., Mosbach, R.: Entrapment of enzymes and microorganisms in synthetic cross-linked polymers and their application in column techniques. Acta chem. scand. **20**, 2807–2810 (1966).

Mueller, H.: Characterization of the molecular region containing the active sites of myosin. J. biol. Chem. **240**, 3816–3828 (1965).

Mueller, H. E.: The preparation of pure antibodies. Z. Immun.-Forsch., Allergie Klin. Immunol. **130**, 164–169 (1966).

Mueller-Eckhardt, C., Luescher, E. F.: Immune reaction of human thrombocytes with γ-globulin-loaded latex particles. Z. ges. inn. Med. **24**, 526–529 (1969).

Nakane, P. K.: Simultaneous localization of multiple tissue antigens using the peroxidase-labeled antibody method. A study on pituitary glands of the rat. J. Histochem. Cytochem. **16**, 557–560 (1968).

Nakane, P. K.: Classifications of anterior pituitary cell types with immunoenzyme histochemistry. J. Histochem. Cytochem. **18** (1), 9–20 (1970).

Nakane, P. K., Nichoalds, G. E., Oxender, D. L.: Cellular localization of leucine-binding protein from *Escherichia coli*. Science **161**, 182–183 (1968).

Nakane, P. K., Pierce, G. B., Jr.: Enzyme-labeled antibodies: preparation and application for the location of antigens. J. Histochem. Cytochem. **14**, 929–931 (1966).

Nakane, P. K., Pierce, G. B., Jr.: Enzyme labeled antibodies for the light and electron microscopic localization of tissue antigens. J. Cell Biol. **33**, 307–318 (1967).

NAYAK, R. V., MCGARRY, E. E., BECK, J. C.: Fluorescein-conjugated antiserum to growth hormones. I. Localization of antiserum in the pituitary in various species. Canad. J. Physiol. Pharmacol. **47**, 15–19 (1969a).

NAYAK, R. V., MCGARRY, E. E., BECK, J. C.: Fluorescein-conjugated antiserum to growth hormones. II. Inhibition studies in the pituitary in various species. Canad. J. Physiol. Pharmacol. **47**, 21–23 (1969b).

NEGORO, H.: Continuous inversion of sucrose by using insoluble saccharose. J. Ferment. Technol. **48** (11), 689–697 (1970).

NEGORO, H., FUKUMOTO, J.: Purification and crystallization of enzymes by forming protein-detergent complexes. Koso Kagaku Shimpojiumu **14**, 334–343 (1960).

NEWCOMB, T. F., HOSHIDA, M.: Factor V and thrombin. Scand. J. clin. Lab. Invest., Suppl. **84**, 61–69 (1965).

NEZLIN, R. S.: Isolation of antibodies endowed with a specific effect against influenza with the aid of influenza virus fixed on cellulose. Dokl. Akad. Nauk SSSR **131**, 676–679 (1960).

NEZLIN, R. S.: Recent methods for the isolation of pure antibodies. Usp. sovrem. Biol. **52**, 19–39 (1961).

NEZLIN, R. S., KULPINA, L. M.: Cellulose-fixed immunoglobulin subunits. Immunochemistry **4**, 269–272 (1967).

NIHEI, T., KAY, C. M.: Isolation and properties of an enzymatically active fragment from papain-digested myosin. Biochim. biophys. Acta (Amst.) **160**, 46–52 (1968).

NISONOFF, A., WISSLER, F. C., LIPMAN, L. N.: The major component of a peptic digest of rabbit antibody. Science **132**, 1770–1771 (1960a).

NISONOFF, A., WISSLER, F. C., LIPMAN, L. N., WOERNLEY, D. L.: Separation of univalent fragments from the bivalent rabbit antibody molecule by reduction of disulfide bonds. Arch. Biochem. **89**, 230–244 (1960b).

NYERGES, G.: Reversed passive hemagglutination-inhibition (HAI) method for the estimation of antivaccinia antibodies. Immunology **17**, 333–337 (1969).

OLOVNIKOV, A. M.: Preparation of immunological adsorbents on the basis of emulsion non-isotactic polystyrene. Lab. Delo **8**, No 8, 31–34 (1962).

OLOVNIKOV, A. M.: Use for immunosorbents for the assay of absolute amounts of haptens and antigens. Biokhimiya **29**, 680–684 (1964a).

OLOVNIKOV, A. M.: Polycondensed antibody immunosorbent and its use in the agglutination reaction for determination of antigen content. Dokl. Akad. Nauk SSSR **158**, 1202–1205 (1964b).

OLOVNIKOV, A. M.: Preparation of immunosorbents as suspensions of diazotized benzidine-protein complexes and their use for the determination of antibodies by means of agglutination. Vopr. med. Khim. **10**, 538–542 (1964c).

OLOVNIKOV, A. M.: Determination of antigen content by agglutination of erythrocytes coated with antiserum proteins polycondensed by tetrazotate-diaminodiphenylamine. Dokl. Akad. Nauk SSSR **169**, 1180–1183 (1966).

OLOVNIKOV, A. M.: Sensitization of erythrocytes by polycondensed proteins of immune serum and their use for determining antigen content. Immunochemistry **4**, 77–80 (1967).

OLOVNIKOV, A. M.: Determination of antigen content. Method for agglutinating PPIS (polycondensed proteins of immune serum)-erythrocytes and method for agglutinating particles of a polycondensed antibody-immunosorbent. Immunokhim. Anal. **1968a**, 67–78.

OLOVNIKOV, A. M.: Methods of immunochemical analysis. Zh. Vses. Khim. Obshchest. **13**, 445–452 (1968b).

OLOVNIKOV, A. M., GURVICH, A. E.: Immunization with protein-cellulose copolymer (immunosorbent). Nature (Lond.) **209**, 417–419 (1966).

OMENN, G. S., ONTJES, D. A., ANFINSEN, C. B.: Fractionation of antibodies against staphylococcal nuclease on Sepharose immunoadsorbents. Nature (Lond.) **225**, 189–190 (1970).

ONG, E. B., TASANG, Y., PERLMANN, G. E.: Action of water-insoluble trypsin derivatives on pepsinogen. J. biol. Chem. **241**, 5661–5666 (1966).

ONKELINX, E., MEULDERMANS, W., JONIAU, M., LONTIE, R.: Glutaraldehyde as a coupling reagent in passive hemagglutination. Immunology **16**, 35–43 (1969).

ONTJES, D. A., ANFINSEN, C. B.: Synthetic studies of structure-function relation in staphylococcal nuclease. Synthetic analogs of fragment P_2. J. biol. Chem. **244** (23), 6316–6322 (1969a).

ONTJES, D. A., ANFINSEN, C. B.: Solid phase synthesis of a 42-residue fragment of staphylococcal nuclease: properties of a semisynthetic enzyme. Proc. nat. Acad. Sci. (Wash.) **64**, 428–435 (1969b).

OPARIN, A. I.: Action of enzymes included in coacervate drops. Proc. Intern. Symposium Enzyme Chem. Tokyo and Kyoto **2**, 433–436 (1957).

OPARIN, A. I., GELMAN, N. S., DEBORIN, G. A.: Significance of structure in certain enzymic processes. Arch. Biochem. **69**, 582–591 (1957).

OPARIN, A. I., SEREBROVSKAYA, K. B.: Action of ribonuclease included in coacervate droplets. Dokl. Akad. Nauk SSSR **122**, 661–664 (1958).

OPARIN, A. I., SEREBROVSKAYA, K. B.: Enzymic processes in coacervate systems. Rev. Roumaine Biochim. **3**, 111–115 (1966).

OPARIN, A. I., SEREBROVSKAYA, K. B., VASIL'EVA, N. V., SAMSONOVA, V. M.: Ribonuclease activity studies in a phosphatide-protein coacervate. Dokl. Akad. Nauk SSSR **179**, 976–978 (1968).

ORESKES, I.: Use of immunoadsorbents for the fractionation of rheumatoid factor and rabbit antihuman γG antibodies. Immunology **11**, 489–498 (1966).

ORLOV, G. E., GURVICH, A. E.: Immunization of animals with a protein-Sephadex complex for the large scale production of antibodies. Vop. med. Khim. **15**, 150–154 (1969).

ORLOV, G. E., YAGLOM, D. L., GURVICH, A. E., VIRNIK, A. D., ROGOVIN, Z. A.: Synthesis of immunosorbents based on cellulose cyanurate, the 3-chloro-2-hydroxypropyl ether of cellulose and a graft copolymer of cellulose and poly(glycidyl methacrylate). Vop. med. Khim. **15**, 287–291 (1969).

PASCOE, P. F.: Autospecific protein complexes with linear cationic polyelectrolyte macromolecules. Kolloid-Z. Z. Polym. **232**, 680–688 (1969).

PENSKY, J., MARSHALL, J. S.: Studies on thyroxine-binding globulin (TBG). II. Separation from human serum by affinity chromatography. Arch. Biochem. **135**, 304–310 (1969).

PERUCCA, P. J., FAULK, W. P., FUDENBERG, H. H.: Passive immune lysis with chromic chloride-treated erythrocytes. J. Immunol. **102**, 812–820 (1969).

PHILLIPS-QUAGLIATA, J. M., LEVINE, B. B., UHR, J. W.: Mechanism of binding of antigen-antibody complexes to phagocytes. Fed. Proc. **28**, 308 (1969a).

PHILLIPS-QUAGLIATA, J. M., LEVINE, B. B., UHR, J. W.: Studies on the mechanism of binding of immune complexes to phagocytes. Nature (Lond.) **222**, 1290–1291 (1969b).

PLESCIA, O. J., BRAUN, W. (eds.): Nucleic acids in immunology. Berlin-Heidelberg-New York: Springer 1968.

POLTORAK, O. M., VOROB'EVA, E. S.: Adsorption models of biological membranes and alkaline phosphatase activation in different types of adsorption layers. Zh. Fiz. Khim. **40**, 1665–1667 (1966).

POLTORAK, O. M., VOROB'EVA, E. S.: Adsorption and catalytic activity of enzymes. Vestn. Mosk. Univ., Ser. II **22** (5), 41–52 (1967).

POLYAKOV, I. I., RASSUDOV, S. M., SHOSHIEV, L. N., LOZOVOI, N. V., SAGATOVSKII, V. N.: Evaluation of various methods of obtaining antigens from Brucellae for preparation of erythrocyte diagnostic agents. Zh. Mikrobiol. (Mosk.) **46**, 77–81 (1969).

PORATH, J.: Molecular sieving and adsorption. Nature (Lond.) **218**, 834–838 (1968).

PORTER, R. R.: Hydrolysis of rabbit γ-globulin and antibodies with crystalline papain. Biochem. J. **73**, 119–126 (1959).

PRISTOUPIL, T. I., FRICOVA, V., HRUBA, A.: Microchromatography, immunochromatography, and detection of proteolytic activity on nitrocellulose membranes impregnated with proteins. J. Chromatog. **26**, 127–131 (1967).

QUABBE, H. J.: Modification of the procedure of Hales and Randle for the radioimmunoassay of insulin. Diabetologia **5**, 101–107 (1969).

RADERMECKER, M., GOODFRIEND, L.: Absorption of human gamma-G globulin by a water-insoluble polymer of specific antiserum. Immunochemistry **6**, 484–488 (1969).

RAO, S. N., HARA, L., ASKARI, A.: Alkali cation-activated AMP deaminase of erythrocytes: some properties of the membrane-bound enzyme. Biochim. biophys. Acta (Amst.) **151**, 651–654 (1968).

Reichlin, M., Schnure, J. J., Vance, V. K.: Induction of antibodies to porcine ACTH in rabbits with nonsteroidogenic polymers of BSA and ACTH. Proc. Soc. exp. Biol. (Wash.) **128**, 347–350 (1968).

Reichnitz, G. A., Llenado, R.: Enzyme electrode for amygdalin. Analyt. Chem. **43**, 283 (1971).

Rejnek, J., Mage, R. G., Reisfeld, R. A.: Rabbit light chains lacking b-allotypic specificities. I. Isolation and characterization of light chains from normal and allotype suppressed homozygotes. J. Immunol. **102**, 638–646 (1969).

Renoux, G., Renoux, M.: Splenic immunocytes specific for bacterial antigens manifested by chromium-induced antigen-erythrocyte linkage. C. R. Acad. Sci. (Paris), Ser. D **268**, 2758–2759 (1969).

Renoux, M., Dubois, M. P., Renoux, G.: Hemagglutination after coupling of bacterial glycoprotein by chromium chloride; failure of benzidine. Ann. Inst. Pasteur **115**, 978 (1968).

Richards, F. F., Pincus, J. H., Bloch, K. J., Barnes, W. T., Haber, E.: The relationship between antigenic complexity and heterogeneity in the antibody response. Biochemistry **8**, 1377–1384 (1969).

Riesel, E., Katchalski, E.: Preparation and properties of water-insoluble derivatives of urease. J. biol. Chem. **239**, 1521–1524 (1964).

Rimon, A., Alexander, B., Katchalski, E.: Action of water-insoluble trypsin derivative on prothrombin and related clotting factors. Biochemistry **5**, 792–798 (1966).

Rimon, A., Gutman, M., Rimon, S.: The activation of plasminogen. I. Preparation and properties of an insoluble derivative of streptokinase. Biochim. biophys. Acta (Amst.) **73**, 301–310 (1963).

Rimon, A., Sela, M.: Chemical modification of human erythrocytes by attachment of tyrosine and alanine peptides. Biochim. biophys. Acta (Amst.) **124**, 408–410 (1966).

Rimon, S., Stupp, Y., Rimon, A.: The activation of plasminogen. III. Insoluble derivatives of the proactivator and the activator. Canad. J. Biochem. **44**, 415–421 (1966).

Robbins, J. B., Haimovich, J., Sela, M.: Purification of antibodies with immunoadsorbents prepared using bromoacetylcellulose. Immunochemistry **4**, 11–22 (1967).

Robinson, P. J., Dunnill, P., Lilly, M. D.: Porous glass as a solid support for immobilisation or affinity chromatography of enzymes. Biochim. biophys. Acta (Amst.) **242**, 659–661 (1971).

Roelants, G. E., Goodman, J. E.: Chemical nature of macrophage RNA-antigen complexes and their relevance to immune induction. J. exp. Med. **130**, 557–574 (1969).

Rohrback, G. H., Scott W. R., Jr.: Biological-chemical sensor for detecting any system affecting life processes. U.S. 3,403,081, 24 Sept. 1968, 5 pp.

Rokhlin, O. V., Nezlin, R. S.: Antigen determination by means of pure antibodies fixed on cellulose. Vop. med. Khim. **15**, 439–443 (1969).

Romeo, D., de Bernard, B.: Formation in vitro of a lipoprotein membrane masking lysozyme activity. Nature (Lond.) **212**, 1491–1492 (1966).

Rothen, A., Mathot, C.: Immunological reactions carried out at a liquid-solid interface with the help of a small electric current. Abstr. Papers, Amer. Chem. Soc., 158th, **1969a**, Coll-95.

Rothen, A., Mathot, C.: Immunoelectroadsorption. Fundamentals of the immunoelectroadsorption method. Immunochemistry **6**, 241–251 (1969b).

Rothen, A., Mathot, C., Thiele, E. H.: Determination of human and bovine growth hormones in the physiological range by immunoelectroadsorption. Experientia (Basel) **25**, 420–421 (1969).

Rothenberg, S. P., Gizis, F., Kamen, B.: Antibodies against folic acid. I. In vitro biophysical effect. J. Lab. clin. Med. **74**, 662–671 (1969).

Rubalcava, B., Martinez de Munoz, D., Gitler, C.: Interaction of fluorescent probes with membranes. I. Effect of ions on erythrocyte membranes. Biochemistry **8**, 2742–2747 (1969).

Salmon, S. E., Mackey, G., Fudenberg, H. H.: "Sandwich" solid phase radioimmunoassay for the quantitative determination of human immunoglobulins. J. Immunol. **103**, 129–137 (1969).

Sander, E. G., McCormick, D. B., Wright, L. D.: Column chromatography of nucleotides over thymidylate-cellulose. J. Chromatog. **21** (3), 419–423 (1966).

Schenck, J. R., Hargie, M. P., Brown, M. S., Ebert, D. S., Yoo, A. L., McIntire, F. C.: Enhancement of antibody formation by *Escherichia coli* lipopolysaccharide and detoxified derivatives. J. Immunol. **102**, 1411–1422 (1969).

Schiff, R., Krieg, R. J., Hunter, R. L.: Localization by peroxidase-labeled antibodies of bovine chymotrypsinogen. J. Histochem. Cytochem. **18** (3), 195–200 (1970).

Schumacher, K., Schneider, W.: Isolation of soluble antigens by specific immunoabsorption. Z. klin. Chem. Klin. Biochem. **7**, 181–182 (1969).

Schwabe, C.: Leucine aminopeptidase. A calcium phosphate gel complex. Biochemistry **8**, 795–802 (1969).

Sehon, A. H.: Chromatographic methods of antibody purification. Symp. Ser. Immunobiol. Stand. **4**, 51–70 (1967).

Seki, T., Jenssen, T. A., Levin, Y., Erdos, E. G.: Active water-insoluble derivative of renin. Nature (Lond.) **225**, 864–865 (1970).

Sela, M., Ungar-Waron, H.: Preparation of nucleoside-specific synthetic antigens. Meth. Enzymol. **12**, Pt. B, 900–903 (1968).

Selegny, E., Avrameas, S., Broun, G., Thomas, D.: Membranes with enzymic activity. Synthesis of membranes with covalently bound enzymes; characterization of the catalytic activity by diffusion-reaction. C. R. Acad. Sci. (Paris), Ser. C **266**, 1431–1434 (1968).

Selegny, E., Broun, G., Geffroy, J., Thomas, D.: Method for determining the true Michaelis constant (K_m) of an enzyme from the stationary state of a membrane with enzyme activity. J. Chim. Phys. Physicochim. Biol. **66**, 391–392 (1969a).

Selegny, E., Broun, G., Thomas, D.: Enzymes in an insoluble phase. Variation of the enzymatic activity as a function of the substrate concentration. Effects of "regulators" on enzymatic membranes. C. R. Acad. Sci. (Paris), Ser. D **269**, 1330–1333 (1969b).

Self, D. A., Kay, G., Lilly, M. D., Dunnill, P.: Conversion of benzyl penicillin to 6-aminopenicillanic acid using an insoluble derivative of penicillin amidase. Biotech. Bioeng. **11**, 337–348 (1969).

Serebrovskaya, K. B.: Enzymic synthesis of polyadenine in coacervates. Probl. Evolyutsionnoi i Tekhn. Biokhim., Akad. Nauk SSSR, Inst. Biokhim. **1964**, 127–130.

Serebrovskaya, K. B., Samsonova, V. M., Minozhedinova, N. S.: Adenosinetriphosphatase activity of myosin in a phosphatide-protein coacervate. Dokl. Akad. Nauk SSSR **180** 1250–1252 (1968).

Serebrovskaya, K. B., Vasil'Eva, N. V.: Conversion of coacervate drops into dynamically stable systems. Dokl. Akad. Nauk SSSR **155**, 212–215 (1964).

Serebrovskaya, K. B., Vasil'Eva, N. V.: Effect of proteins on the activity of ribonuclease in oleate coacervate. Dokl. Acad. Nauk SSSR **167**, 457–460 (1966).

Shainoff, J. R.: A cellulose mercurial: preparation, properties, and use in antibody purification. J. Immunol. **100**, 187–193 (1968).

Shaltiel, S., Mizrahi, R., Stupp, Y., Sela, M.: Reduction in immunological manifestations of 6-aminopenicillanic acid by treatment with water insoluble pronase. Europ. J. Biochem. **14**, 509–515 (1970).

Sharp, A. K., Kay, G., Lilly, M. D.: Kinetics of β-galactosidase attached to porous cellulose sheets. Biotech. Bioeng. **11**, 363–380 (1969).

Shimazaki, M., Hamba, K., Hiramine, C., Katsu, N., Mori, T., Keno, Y.: An attempt to purify anti-thyroglobulin antibodies by means of the immunoadsorbents. Wakayama Med. Rep. **11**, 59–66 (1966).

Silman, H. I., Cebra, J. J., Givol, D.: Carboxyl-terminal amino acids of rabbit γ-globulin. J. biol. Chem. **237**, 2196–2200 (1962).

Silman, H. I., Karlin, A.: Effect of local pH changes caused by substrate hydrolysis on the activity of membrane-bound acetylcholinesterase. Proc. nat. Acad. Sci. (Wash.) **58**, 1664–1668 (1967).

Silman, I. H.: Microenvironmental effects on enzyme activity. J. gen. Physiol. **54**, S 50–S 57 (1969).

Silman, I. H., Katchalski, E.: Water-insoluble derivatives of enzymes, antigens, and antibodies. Ann. Rev. Biochem. **35**, 873–908 (1966).

Simonian, S. J., Gill, T. J., III., Gershoff, S. N.: Synthetic polypeptide antigens. XX. Genetic control of the antibody response in the rat to structurally different synthetic polypeptide antigens. J. Immunol. **101**, 730–742 (1968).

SLOBIN, L. I.: Antigenic properties of a homogeneous tobacco mosaic virus-hapten conjugate. Nature (Lond.) **225**, 698–701 (1970).

SLUYTERMAN, L. A. A., DE GRAAF, M. J. M.: Activity of papain in crystalline state. Biochim. biophys. Acta (Amst.) **171**, 277–287 (1969).

SLUYTERMAN, L. A. A., WIJDENES, J.: An agarose mercurial column for the separation of mercaptopapain and nonmercaptopapain. Biochim. biophys. Acta (Amst.) **200**, 593–595 (1970).

SORG, C., HIMMELSPACH, K., RUEDE, E., WESTPHAL, O.: Synthesis of stearoaminoethyl-(1-iminooctadecylaminoethyl)poly(vinyl alcohol) and its use for preparation of agents for specific adsorptive sensitization of erythrocytes. Israel J. med. Sci. **5**, 253–258 (1969).

SPARKS, R. E., SALEMME, R. M., MEIER, P. M., LITT, M. H., LINDAN, O.: Removal of waste metabolites in uremia by microencapsulated reactants. Trans. Amer. Soc. Artificial Organs **15**, 353 (1969).

SPITLER, L., HUBER, H., FUDENBERG, H. H.: Inhibition of capillary migration by antigen-antibody complexes. J. Immunol. **102**, 404–411 (1969).

SPITZER, R. H., KAPLAN, M. A., LEIJA, J. G.: Functional activities of γ-G ragweed antibodies isolated by an immunospecific method. Int. Arch. Allergy **34**, 488–494 (1968).

SPRAGG, J., TALAMO, R. C., SUZUKI, K., APPELBAUM, D. M., AUSTEN, K. F., HABER, E.: Structural requirements for binding of bradykinin to antibody. II. Studies with bradykinin fragments. Biochemistry **7**, 4086–4089 (1968).

SPROSSLER, B., LINGENS, F.: Affinity chromatography of the chorismate mutase from *Claviceps paspali* based on specific interaction with the activator. FEBS (Fed. Eur. Biochem. Soc.) Letters **6** (3), 232–234 (1970).

STEINBUCH, M., MATTE, C., AUDRAN, R., MOULLEC, J.: Molecular support of rheumatoid factor and anti-Gm (a) activity. Nature (Lond.) **207**, 92 (1965).

STEINBUCH, M., PEJAUDIER, L.: Activation of plasminogen by a streptokinase coupled to a cellulose support. Bibl. haemat. (Basel) No 19, 169–173 (1962).

STEPHEN, J., GALLOP, R. G. C., SMITH, H.: Separation of antigens by immunological specificity. Use of disulfide-linked antibodies as immunosorbents. Biochem. J. **101**, 717–720 (1966).

STEPHEN, J., SMITH, H.: Disulfide-linked antibodies as immunosorbents. Biochem. J. **94**, 23P (1965).

SUNDARAM, P. V., HORNBY, W. E.: Preparation and properties of urease chemically attached to nylon tube. FEBS (Fed. Eur. Biochem. Soc.) Letters **10** (5), 325 (1970).

SUNDARAM, P. V., TWEEDALE, A., LAIDLER, K. J.: Kinetic laws for solid-supported enzymes. Canad. J. Chem. **48**, 1498–1504 (1970).

SURINOV, B. P., KASHKIN, K. P., BOCHKOVA, D. N., LORENTS, V. G.: Use of an immunosorbent for preparing tissue antibodies. Bull. exp. Biol. Med. USSR **65**, 703 (1968).

SUTHERLAND, G. B.: Immunological technique for protein isolation. Science **132**, 1252–1253 (1960).

SUTHERLAND, G. B., DATTAGUPTA, U.: Isolation of tumor specific antigen by an immunological method. Fed. Proc. **28** , 699 (1969).

SUTHERLAND, G. B., MCDUFFIE, N. M.: Immunochemistry of coagulated antigen-antibody complex. II. Parameters affecting the adsorption of native antigen by the complex and its subsequent desorption. J. Immunol. **90**, 444–447 (1963).

SUZUKI, H., OZAWA, Y., MAEDA, H.: Water-insoluble enzyme. Hydrolysis of sucrose by insoluble yeast invertase. Agr. Biol. Chem. (Tokyo) **30**, 807–812 (1966).

SUZUKI, H., OZAWA, Y., MAEDA, H., TANABE, O.: Water insoluble enzyme. Hydrolysis of sucrose by insoluble yeast invertase. Kogyo Gijutsuin, Hakko Kenkyusho Kenkyu Hokoku No 31, 11–19 (1967).

SWOPE, A. E., KAHN, R. H., CONKLIN, J. L.: A comparison of Alcian blue-aldehyde fuchsin and peroxidase-labeled antibody staining techniques in the rat adenohypophysis. J. Histochem. Cytochem. **18** (6), 450–454 (1970).

TAKAHASHI, M., YAGI, Y., PRESSMAN, D.: Preparation of fluorescent antibody reagents monospecific to heavy chains of human immunoglobulin. J. Immunol. **100**, 1169–1175 (1968).

TAKAHASHI, M., YAGI, Y., PRESSMAN, D.: Preparation of fluorescent antibody reagents monospecific to light chains of human immunoglobulin. J. Immunol. **102**, 1268–1273 (1969).

Talamo, R. C., Haber, E., Austen, K. F.: Antibody to bradykinin: effect of carrier and method of coupling on specificity and affinity. J. Immunol. **101**, 333–341 (1968).
Tallberg, T.: A new agglutination method using small insoluble antibody- or antigen-polymer particles. Ann. Med. exp. Fenn. **45**, 477–486 (1967).
Tallberg, T.: A simple method of preparing specific human anti-tetanus toxoid gamma globulins using immunoadsorbents. Scand. J. clin. Lab. Invest. **23**, Suppl. 108, 49 (1969).
Tallberg, T., Eskola, L.: Slide agglutination using antibody polymer particles immunologically coated with antigen. Scand. J. clin. Lab. Invest. **23**, Suppl. 108, 50 (1969).
Tallberg, T., Eskola, L., Linturi, T. I.: Immunoassay of antigenic hormones using antibody-active particles. Scand. J. clin. Lab. Invest. **24**, Suppl. 110, 67 (1969).
Tallberg, T., Jormalainen, S. A.: Studies on phylogeny of human immunoglobulins and preparation of specific anti-globulin sera using immunoadsorbents. Scand. J. clin. Lab. Invest. **21**, Suppl. 101, 35 (1968).
Tallberg, T., Kaivola, S., Ehnholm, C.: Immunological purification of human protein hormones using antibody-active polymers. Scand. J. clin. Lab. Invest. **21**, Suppl 101, 37 (1968a).
Tallberg, T., Linder, E., Hjelt, L.: Further studies on a human thymus specific protein. I. Immunological purification of the thymus specific antigen lacking in patients with lymphatic leukemia and the detection of an ileo-jejunal protein antigen in the thymus. Scand. J. clin. Lab. Invest. **21**, Suppl. 101, 36 (1968b).
Tallberg, T., Linturi, T. I.: A simple radioimmune assay technique for human serum insulin using biologically active, insoluble anti-insulin serum polymers. Scand. J. clin. Lab. Invest. **21**, Suppl. 101, 37 (1968).
Taniuchi, H., Anfinsen, C. B.: An experimental approach to the study of the folding of staphylococcal nuclease. J. biol. Chem. **244** (14), 3864–3875 (1969).
Thorell, J. I.: Use of immunosorbent in the preparation of an "antigen-free" plasma. An adequate diluent for standards in radio-immunoassay. Clin. chim. Acta **22**, 579–582 (1968).
Timpl, R., Furthmayr, H., Steffen, C., Doleschel, W.: Isolation of pure anti-collagen antibodies using a specific immunoadsorbent technique. Z. Immun.-Forsch. Allerg. Klin. Immunol. **134**, 391–396 (1967).
Timpl, R., Wolff, I., Furthmayr, H., Steffen, C.: Immunogenicity and specificity of collagen. VI. Separation of antibody fractions with restricted specificity from anti-collagen sera by using an immunoadsorbent technique. Immunology **15**, 145–151 (1968).
Torrigiani, G., Roitt, I. M.: Estimation of antibodies in different immunoglobulin clases. J. Immunol. **102**, 492–495 (1969).
Tosa, T., Chibata, I.: Properties and uses of water insoluble enzymes. Hakko Kyokaishi **25** (2), 49–57 (1967).
Tosa, T., Mori, T., Chibata, I.: Studies on continuous enzyme reactions. VI. Enzymatic properties of the DEAE-Sephadex-aminoacylase complex. Agr. Biol. Chem. **33**, 1053–1059 (1969a).
Tosa, T., Mori, T., Fuse, N., Chibata, I.: Continuous enzyme reactions. I. Screening of carriers for preparation of water-insoluble aminoacylase. Enzymologia **31**, 214–224 (1966a).
Tosa, T., Mori, T., Fuse, N., Chibata, I.: Continuous enzyme reactions. II. Preparation of diethylamino-ethylcellulose-aminoacylase column and continuous optical resolution of acetyl-DL-methionine. Enzymologia **31**, 225–238 (1966b).
Tosa, T., Mori, T., Fuse, N., Chibata, I.: Continuous enzyme reactions. IV. Preparation of a DEAE-Sephadex-aminoacylase column and continuous optical resolution of acyl-DL-amino acids. Biotech. Bioeng. **9**, 603–615 (1967a).
Tosa, T., Mori, T., Fuse, N., Chibata, I.: Studies on continuous enzyme reactions. III. Enzymic properties of the DEAE-cellulose-aminoacylase complex. Enzymologia **32**, 153–168 (1967b).
Tosa, T., Mori, T., Fuse, N., Chibata, I.: Studies on continuous enzyme reactions. V. Kinetics and industrial application of aminoacylase columns for continuous optical resolution of acyl-DL-amino acids. Agr. Biol. Chem. **33**, 1047–1052 (1969b).
Traub, A., Kaufmann, E., Teitz, Y.: Synthesis of nicotinamide adenine dinucleotide by NAD pyrophosphorylase on a column of hydroxylapatite. Analyt. Biochem. **28**, 469–476 (1969).

Tribouley, J., Tribouley-Duret, J., Irribarren, M. J., Pautrizel, R.: Evidence of specific antibodies for *Fasciola hepatica* by passive hemagglutination using glutaraldehyde as a coupling agent. C. R. Acad. Sci. (Paris), Ser. D **268**, 2215–2217 (1969).

Truffa-Bachi, P., Wofsy, L.: Specific separation of cells on affinity columns. Proc. nat. Acad. Sci. (Wash.) **66** (3), 685–692 (1970).

Tsumura, N., Ishikawa, M.: Continuous isomerization of glucose by a column of glucose isomerase. Nippon Shokuhin Kogyo Gakkaishi **14**, 539–540 (1967).

Ungar-Waron, H., Hurwitz, E., Jaton, J. C., Sela, M.: Antibodies elicited with conjugates of nucleosides with synthetic polypeptides. Biochim. biophys. Acta (Amst.) **138**, 513–531 (1967).

Updike, S. J., Harris, D. R., Shrago, E.: Microorganisms, alive and imprisoned in a polymer cage. Nature (Lond.) **224**, 1122 (1969).

Updike, S. J., Hicks, G. P.: The enzyme electrode. Nature (Lond.) **214**, 986–988 (1967a).

Updike, S. J., Hicks, G. P.: Reagentless substrate analysis with immobilized enzymes. Science **158**, 270–272 (1967b).

Van der Ploeg, M., Van Duijn, P.: Cytophotometric determination of alkaline phosphatase activity of individual neutrophilic leukocytes with a biochemically calibrated model system. J. Histochem. Cytochem. **16**, 693–706 (1968).

Van Duijn, P., Pascoe, E., Van der Ploeg, M.: Theoretical and experimental aspects of enzyme determination in a cytochemical model system of polyacrylamide films containing alkaline phosphatase. J. Histochem. Cytochem. **15**, 631–645 (1967).

Vannier, W. E., Bryan, W. P., Campbell, D. H.: The preparation and properties of a hapten-cellulose antibody adsorbent. Immunochemistry **2**, 1–12 (1965).

Vardanis, A.: Glycogen-bound enzymes: a new method of isolation. Arch. Biochem. **130**, 408–412 (1969).

Vicari, G., Collotti, C., Quattrocchi, R., Scafti, A. R.: Detection of the cellular immune response against bacteriophage ΦX174 by an adaptation of the hemolytic plaque technique. Immunology **15**, 661–670 (1968).

Vorob'eva, E. S., Poltorak, O. M.: Comparative study of enzymic activity of acid phosphatase and phosphoglucomutase in adsorption layers of various types. Vestn. Mosk. Univ., Ser. II **21** (6), 17–20 (1966a).

Vorob'eva, E. S., Poltorak, O. M.: Deactivation of hexokinase in adsorption layers of various types. Zh. Fiz. Khim. **40**, 2596–2598 (1966b).

Voss, E. W., Jr., Corley, K., Huang, L. H.: Anti-hapten antibody induction with deoxyribonucleic acid carrier. Immunochemistry **6**, 361–370 (1969).

Vreeland, V.: Localization of a cell wall polysaccharide in a brown alga with labeled antibody. J. Histochem. Cytochem. **18**, 371 (1970).

Vyas, G. N., Fudenberg, H. H., Pretty, H. M., Gold, E. R.: A new rapid method for genetic typing of human immunoglobulins. J. Immunol. **100**, 274–279 (1968).

Wagner, T. E., Chai, H. G., Warfield, A. S.: Cellulose polymer supported sequential analysis of polyribonucleotides. J. Amer. chem. Soc. **91**, 2388–2389 (1969).

Walker, W. S., Fishman, M.: *In vitro* induction of phage antibody and its detection by passive hemagglutination. Fed. Proc. **28**, 427 (1969).

Walker, W. S., Liu, C. T., Adler, F. L.: Detection of anti-phage antibody by passive hemagglutination. J. Immunol. **103**, 907–916 (1969).

Weetall, H. H.: Immunoadsorbent for the isolation of bacterial specific antibodies. J. Bact. **93**, 1876–1880 (1967).

Weetall, H. H., Baum, G.: Continuous processing of DL-amino acids with an insolubilized L-amino acid oxidase. Abstr. Papers Amer. Chem. Soc., 158th, Sept. **1969**, Biol. 153.

Weetall, H. H., Weliky, N.: A new technique for the enzymic detection of hydrogen peroxide Analyt. Biochem. **14**, 160–162 (1966).

Weibel, M. K., Bright, H. J.: Insolubilized enzymes. Kinetic behavior of glucose oxidase bound to porous glass particles. Biochem. J. **124**, 801–807 (1971).

Weinbach, R.: Preparation and use of glutaraldehyde-fixed erythrocytes. Fed. Proc. **28**, 316 (1969).

Weliky, N., Weetall, H. H.: Chemistry and use of cellulose derivatives for the study of biological systems. Immunochemistry **2**, 293–322 (1965).

Westman, T. L.: A soluble, stable polyelectrolyte derivative of trypsin. Biochem. biophys. Res. Commun. **35**, 313–317 (1969).

Wharton, C. W., Crook, E. M., Brocklehurst, K.: Nature of the perturbation of the Michaelis constant of the bromelain-catalyzed hydrolysis of α-N-benzoyl-L-arginine ethyl ester consequent upon attachment of bromelain to O-(carboxyl-methyl)cellulose. Europ. J. Biochem. **6**, 572–578 (1968).

Wheeler, K. P., Edwards, B. A., Whittam, R.: Properties of two phosphatases attached to insoluble cellulose matrices. Biochim. biophys. Acta (Amst.) **191**, 187–189 (1969).

Whittam, R., Edwards, B. A., Wheeler, K. P.: An approach to the study of enzyme action in artificial membranes. Biochem. J. **107**, 3P–5P (1968).

Wicker, R., Avrameas, S.: Localization of virus antigens by enzyme-labelled antibodies. J. gen. Virol. **4**, 465–471 (1969).

Wide, L.: A radio-immunological method for determination of proteins with use of sephadex-coupled antibodies. Acta pathol. microbiol. scand. **70**, 635 (1967).

Wide, L., Axen, R., Porath, J.: Radioimmunosorbent assay for proteins. Chemical couplings of antibodies to insoluble dextran. Immunochemistry **4**, 381–386 (1967).

Wide, L., Porath, J.: Radioimmunoassay of proteins with the use of Sephadex coupled antibodies. Biochim. biophys. Acta (Amst.) **130**, 257–260 (1966).

Wigzell, H., Andersson, B.: Cell separation on antigen-coated columns. Elimination of high rate antibody-forming cells and immunological memory cells. J. exp. Med. **129**, 23–36 (1969).

Wilchek, M.: Chromatography for the purification of labeled peptides from staphylococcal nuclease. FEBS (Fed. Europ. Biochem. Soc.) Letters **7** (2), 161–163 (1970).

Wilchek, M., Gorecki, M.: Affinity chromatography of bovine pancreatic ribonuclease A. Europ. J. Biochem. **11** (3), 491–494 (1969).

Wilson, R. J. H., Kay, G., Lilly, M. D.: Preparation and kinetics of lactate dehydrogenase attached to water-insoluble particles and sheets. Biochem. J. **108**, 845–853 (1968a).

Wilson, R. J. H., Kay, G., Lilly, M. D.: The preparation and properties of pyruvate kinase attached to porous sheets, and the operation of a two-enzyme continuous-feed reactor. Biochem. J. **109**, 137–141 (1968b).

Wilson, R. J. H., Lilly, M. D.: Preparation and use of insolubilized amyloglucosisdae for production of sweet glucose liquors. Biotech. Bioeng. **11**, 349–362 (1969).

Wofsy, L., Burr, B.: Use of affinity chromatography for specific purification of antibodies and antigens. J. Immunol. **103**, 380–382 (1969).

Wojtyla, B., Lachowicz, T. M., Reiss, J., Zyzynski, Z.: Immunofluorescence adsorption test (IFAT) for titrating tetanus anatoxins. Arch. Immunol. Ther. exp. (Warsz.) **17**, 194–199 (1969).

Wolberg, G., Liu, C.-T., Adler, F. L.: Agglutination of bovine serum albumin coated tanned red cells. Fed. Proc. **28**, 308 (1969a).

Wolberg, G., Liu, C.-T., Adler, F. L.: Passive hemagglutination. I. Titration of early and late antibodies with tanned red cells sensitized with native or denatured bovine serum albumin. J. Immunol. **103**, 879–890 (1969b).

Wolf, B., Izenberg, H.: Purification of two protein antigens by dissociation from thiolated antibody. J. Immunol. **102**, 123–130 (1969).

Wood, K. R., Stephen, J., Smith, H.: Specific adsorption of Semliki forest virus on a disulphide-linked immunosorbent derived from rabbit antibody. J. gen. Virol. **2**, 313–316 (1968).

Yde, H.: Immunoassay of plasma growth hormone. Scand. J. clin. Invest. **24**, Suppl. 110, 68–72 (1969).

Yip, C. C., Logothetopoulos, J.: Specific antiproinsulin serum and the presence of proinsulin in calf serum. Proc. nat. Acad. Sci. (Wash.) **62**, 415–419 (1969).

Yuasa, S., Matsuzaki, K., Nozu, K.: Ribosome-bound RNA polymerase of *Escherichia coli*. Biochim. biophys. Acta (Amst.) **145**, 637–650 (1967).

Zimmerman, R. A., Mathews, J., Wilson, E.: Microtiter indirect hemagglutination procedure for identification of streptococcal M-protein antibodies. Appl. Microbiol. **16**, 1640–1645 (1968).

Zschocke, R. H., Grieble, H. G., Bach, G. L., Anderson, T. O.: Studies on IgA. I. Fractionation procedure for isolation of IgA from pooled normal human plasma. J. Immunol. **102**, 625–637 (1969).

Virus Sensitization

Mirko Majer[1]

With 3 Figures

Contents

I. Introduction

A. Definition and Terminology

The interaction of virus with antibody may result in neutralization (Sternberg, 1982), delayed eclipsing (Andrewes and Elford, 1933a; Wadell, 1970) or it may be without direct effect on virus infectivity. Virus sensitization is a specific attachment of antibody to the virion without concomitant neutralization (Notkins et al., 1966; Notkins, 1971). The resulting infectious virus-antibody complexes can be inactivated by homologous antiglobulin which is inactive against the virus itself. Antibodies directed against other serum components, like albumin, do not possess such inactivating activity (Hahon, 1970). The antiviral antibody is said to "sensitize" the virus to the antiglobulin. Accordingly, the virus-antibody complexes are called "sensitized" virus. By analogy with the "sensitized red cell" the term "sensitization" was also used for interaction of herpes simplex virus with complement requiring neutralizing antibodies (Yoshino and Taniguchi, 1965, 1965a; Daniels et al., 1970). Thus, virus sensitization is a binary antigen-antibody reaction which can be recognized only by the action of a third component, the antiglobulin. The interaction with the antiglobulin results in virus neutralization, a ternary reaction which implies the host cell as an additional component (see review Fazakas

1 Swiss Serum and Vaccine Institute Berne, Berne, Switzerland.

DE ST. GROTH, 1962). The combined effect of the antiviral antibodies and the antiglobulin has been called "indirect neutralization assay", "amplification method" (STEMKE and LENNOX, 1967), "sandwich technique" (KJELLÉN and PEREIRA, 1968) and "neutralization enhancement" (WADELL, 1970).

In an analogous way antiglobulin is used in some other serological systems. The Coombs-test (COOMBS et al., 1945), the enhancement of mycoplasma agglutination titres by use of antiglobulin (ADLER and DAMASSA, 1964) and the indirect fluorescent antibody technique (WELLER and COONS, 1954) are a few examples.

B. Historical Background

The studies on virus sensitization originate from the investigation of the non-neutralizable fraction of viruses (MANDEL, 1958; NOTKINS et al., 1966). ANDREWES and ELFORD (1933) and BURNET et al. (1937) observed first that bacteriophages and animal viruses respectively could not be neutralized completely by antiserum. The residual infectivity of viruses observed after interaction with excess antibodies is called non-neutralizable or persistent fraction. Since the general introduction of tissue culture techniques in virology in the early fifties, the neutralization of animal viruses has been studied in extenso. The existence of the non-neutralizable fraction was described with a variety of viruses like picornaviruses (LEDINKO and MELNICK, 1953; MELNICK, 1954; BRADISH et al., 1962), reoviruses (WALLIS and MELNICK, 1967), group A arboviruses (DULBECCO et al., 1956), group B arboviruses (HASHIMOTO and PRICE, 1963; WESTAWAY, 1965), myxoviruses (RUBIN and FRANKLIN, 1957; LAFFERTY, 1963), adenoviruses (KJELLÉN, 1965), herpesviruses (YOSHINO and TANIGUCHI, 1965) and poxviruses (BURNET et al., 1937; LAFFERTY, 1963). Most of the studies were performed in vitro. However the persistent fraction may occur in vivo, as in lactic dehydrogenase virus infection of mice (NOTKINS et al., 1966), H-1 virus infection of hamster (TOOLAN, 1965) and Visna virus infection of sheep (GUDNADÓTTIR and PÁLSSON, 1965).

The size of the non-neutralizable fraction depends on all three components of the ternary reaction: the virus, the antibody and the host cell. Different strains of the same virus may vary in the size of their non-neutralizable fractions (ITOH and MELNICK, 1957; HAHON, 1970). The donor species of the antiserum (LAFFERTY, 1963) and the antibody class may affect the size of the persistent fraction as well. IgM antibodies leave a larger persistent fraction than IgG antibodies (PHILIPSON, 1966; SVEHAG, 1966; WESTAWAY, 1968). With certain virus-cell systems the influence of the substrate on the non-neutralizable fraction was described (FRIEDMAN, 1954; KJELLÉN and SCHLESINGER, 1959; LAFFERTY, 1963), whereas with some other viruses no cell dependent persistent fraction could be demonstrated (DULBECCO et al., 1956; HAHON, 1970).

Several factors may be responsible for the formation of the non-neutralizable fraction. The existence of a genetic variant could be ruled out. The progeny of non-neutralized virus behaved in the neutralization test in the same manner as the parent population (Dulbecco et al., 1956; Notkins et al., 1966; Ashe and Notkins, 1967; Hahon, 1970; Majer and Link, 1970). The existence of a phenotypic variant seems unlikely as well (Lafferty, 1963; Notkins et al., 1966). Today it is generally accepted that the persistent fraction consists of virus-antibody complexes and not of particles which are by nature unable to react with antibodies, as was once proposed by Dulbecco et al. (1956) (Mandel, 1958; Lafferty, 1963, 1963a; Kjellén, 1965; Notkins et al., 1966; Philipson, 1966). However, the residual infectivity may be caused by virus hindered from contact with antibodies by cell debris (Hampar, 1966) or by virus aggregation (Rappaport, 1957; Wallis and Melnick, 1967; Hahon, 1970). The virus-antibody complexes of the persistent fraction contain sensitized virus (Mandel, 1958; Notkins et al., 1966; Iwasaki and Ogura 1968; Hahon, 1970; Majer and Link, 1970). However, sensitized virus is not restricted to the non-neutralizable fraction and sensitization is now mainly studied to elucidate the mechanism of the neutralization reaction. Furthermore, it represents a sensitive serological system.

II. Sensitization Reaction

A. Assay for Virus Sensitization

Virus sensitization can be recognized only by the action of antiglobulin. Infectious virus-antibody complexes can be identified with physico-chemical methods as well (Kjellén, 1965; Philipson, 1966; Notkins et al., 1971). However, the application of such methods would require the still missing evidence of *a priori* sensitized virus in the complexes. The usual experimental model consists of two stages. In the first stage the virus is incubated with antiviral antibodies (sensitization reaction) and in the second stage the homologous antiglobulin is added (assay reaction). At both stages samples are taken and tested for infectivity.

Antiglobulin should be used in excess, since unbound gammaglobulin in the reaction mixture can prevent the antiglobulin from reacting with the sensitized virus (Ashe and Notkins, 1966). The effect of unbound gammaglobulin was demonstrated experimentally: lactic dehydrogenase virus sensitized with mouse antiserum was mixed with normal whole mouse serum, mouse gammaglobulin and mouse albumin respectively. Gammaglobulin and whole mouse serum inhibited the neutralization of the sensitized virus by antiglobulin, whereas albumin had no inhibitory effect (Notkins et al., 1968). The inhibitory effect of normal serum was confirmed with herpes virus sensitized with human antiserum (Daniels et al., 1970). To guarantee the excess of antiglobulin, the concentration of antivirus antibodies has to be limited (Notkins et al., 1968). If high concentrations of *anti*virus *anti*bodies are required in the sensitization reaction, the sensitized virus may be appropriately diluted prior to the addition of antiglobulin. Furthermore sensitized virus can be separated from unbound globulin by differential centrifugation (Daniels et al., 1970a; Notkins et al., 1971).

In addition to the quantitative relations of the reactants, the above mentioned sequence in which they react with each other is essential. If the virus is inoculated with the antiviral antibodies and the antiglobulin simultaneously or the antibody preparations are mixed and incubated prior to the addition of the virus, virus sensitization may not be detected (MAJER and LINK, 1968; 1970). In these instances both the assay and the sensitization reactions seem to be effected since it was demonstrated that the antiglobulin interferes also with the neutralizing activity of the antiviral antibodies (DUDLEY et al., 1970). For special objects, modifications of the described experimental design have been employed. The sensitized virus may be adsorbed to the host cell prior to the addition of antiglobulin thus excluding immunoaggregation of sensitized virus particles (MANDEL, 1958; WADELL, 1970).

The incubation time and temperature of the assay reaction seem not to be critical. For convenience 4° C, room temperature and 37° C have been used; the time of incubation ranging from 1 to 120 minutes (ASHE and NOTKINS, 1967; KELLER and DWYER, 1968; KJELLÉN and PEREIRA, 1968; HAHON, 1970; MAJER and LINK, 1970; WEEKE-LÜTTMANN, 1970).

In quantitative studies it should be taken into consideration that other virus-antibody interactions like neutralization and aggregation which effect the virus titre may occur. To express the extent of virus sensitization the "neutralization index" was used. This is calculated by subtracting the log of virus titre remaining after the assay reaction from the log of the titre remaining after the sensitization reaction (ASHE and NOTKINS, 1966; NOTKINS et al., 1966). In kinetic studies the ratio of the "amplified neutralization constant" and the "normal neutralization constant" was calculated by DUDLEY et al. (1970). A further method to express the extent of virus sensitization is the "neutralization enhancement index" (NE index) which represents the ratio of the anti-virus serum titre in the assay reaction over the corresponding titre in the sensitization reaction (WADELL, 1970).

The test for infectivity mainly depends on the virus under study. The methods employed include *in vivo* tests (NOTKINS et al., 1966), plaque tests in tissue culture monolayers (ASHE and NOTKINS, 1966), immuno-fluorescent cell counting procedures (HAHON, 1970) and pock counting on chorioallantoic membranes of embryonated eggs (MAJER and LINK, 1970).

B. Virus Suspension

In addition to the virus-determined biological properties of single virus particles, certain characteristics of the virus suspension used for immunological studies may influence the outcome of virus-antibody interactions. In relation to virus sensitization the degree of dispersion and the concentration of the virus are of particular interest.

Virus suspensions usually consist of a mixture of both single virus particles and virus aggregates. Marked differences have been found between the neutralization of monodispersed virus and virus aggregates (WALLIS and MELNICK,

1967; MAJER and LINK, 1970). Although such differences have not been described in the case of the sensitization reaction, monodispersed virus was used in sensitization studies (ASHE et al., 1969; MAJER and LINK, 1970; WALLIS and MELNICK, 1970). The method applied to obtain monodispersed suspensions was the filtration of the native virus preparations through membrane filters of porosity about 2–4 times greater than the diameter of the virus under test (WALLIS and MELNICK, 1970).

The postulated lattice formation of antigen and antibody in the zone of equivalence by MARRACK (1938) and PAULING et al. (1944) was found to be valid for virus-antibody mixtures (ALMEIDA et al., 1963; SMITH et al., 1965; NORRBY et al., 1969). Lattice formation or immuno-aggregation is not necessarily associated with neutralization of virus infectivity, however it may mimic its neutralization in certain test systems (WADELL, 1970). The concentration of the virus suspension will influence the degree of immuno-aggregation. There is a low probability of lattice formation in highly diluted virus suspensions. Arbitrarily, virus preparations may be divided into low concentration suspensions leading to practically no immuno-aggregation, and into high concentration suspensions which may be aggregated readily. No firm limitation can be given so far. WALLIS and MELNICK (1970) used successfully 250 pfu/0.1 ml of herpes simplex virus as a low concentration suspension in the above mentioned sense.

C. Antiviral Antibodies

A variety of viruses have a more or less complex antigenic composition (ANDREWES, 1964). Antibodies directed against different antigens of the virus surface may differ in their ability to neutralize virus infectivity (WEBSTER and LAVER, 1967; KJELLÉN and PEREIRA, 1968). These observations lead to the question which antibody (or antibodies) are involved in the sensitization of complex viruses. So far most studies on virus sensitization have been performed with antisera against whole viria. However, to determine the responsible antibodies, antisera against purified subunits are required. Limited conclusions are also possible from studies with heterotypic antisera. Sera against adenoviruses type 1 and type 2 sensitize adenovirus type 5 without any cross-neutralization (KJELLÉN and PEREIRA, 1968). Further evidence in favour of the concept that sensitizing antibodies are not necessarily identical with neutralizing antibodies was supported by the use of antisera against purified adenovirus components. Anti-fiber antibodies and antibodies against two populations of the hexons—anti-Ac1 and anti-Ac2 antibodies—of adenovirus type 2 were examined (WADELL, 1969; 1970). Only anti-Ac1 antibodies possess neutralizing activity. On the other hand anti-Ac2 and anti-fiber antibodies sensitize the virus to a much greater extent (NE indices: 420 and >1000 respectively) than the neutralizing anti-Ac1 antibodies (NE index: 1,9). Antineuraminidase does not neutralize the infectivity of influenza viruses in the classical neutralization test, whereas anti-haemagglutinin antibodies do

(WEBSTER and LAVER, 1967). Non-neutralizing anti-neuraminidase antiserum sensitized the homologous A2/Singapore/1/57 virus (MAJER and LINK, 1971). However, an influenza virus recombinant containing neuraminidase from its A2/Singapore/1/57 parent (SCHILD et al., 1970) could not be sensitized by the antineuraminidase antiserum (MAJER and LINK, 1972) which may be explained by different structural relationship of the neuraminidase to hemagglutinin in the recombinant strain.

The sensitizing property of antibodies belonging to different classes of immunoglobulins has been studied by several authors. It can be conclusively said that IgG antibodies sensitize viruses. IWASAKI and OGURA (1968) used purified mouse anti-Japanese Encephalitis Virus (anti-JEV) IgG globulin for virus sensitization and rabbit anti-mouse IgG globulin in the assay reaction. Herpes simplex virus was sensitized by purified homologous rabbit IgG immunoglobulin to the goat anti-rabbit serum (HAMPAR et al., 1968). Bacteriophages T2 and f2 have been shown to be sensitized after interaction with IgG antibodies (STEMKE and LENNOX, 1967; DUDLEY et al., 1970). Further evidence was supplied by experiments with unfractionated antivirus sera and corresponding anti-IgG antibodies (NOTKINS et al., 1966; 1968; ASHE et al., 1969; HAHON, 1970; MAJER and LINK, 1970; WALLIS and MELNICK, 1970). The latter approach was applied to studies with other classes of gammaglobulins. Anti-mouse IgF and anti-mouse IgA immunoglobulins neutralized the sensitized lactic dehydrogenase virus, whereas anti-IgM had no such effect (NOTKINS et al., 1968). Analogous results have been achieved with Venezuelan equine encephalitis virus using human antiserum (HAHON, 1970). Anti-human IgA serum but not anti-IgM serum neutralized the sensitized virus. Using purified anti-herpes IgM antibodies, HAMPAR et al. (1968) could sensitize the virus with IgM globulin from "late" sera, whereas no sensitization was achieved with IgM globulin from "early" sera. A presumed correlation between the complement requirement of anti-herpes antibodies and their sensitizing properties could not be established (HAMPAR et al., 1968). Secretory IgA immunoglobulins have been reported to have sensitizing activity as well. Stool extracts of immunized volunteers sensitized poliovirus to anti-human IgA globulin (KELLER and DWYER, 1968).

In the above reported studies untreated bivalent antibodies have been used to sensitize viruses. Univalent Fab fragments prepared by digestion of IgG immunoglobulin with papain (PORTER, 1959) were likewise able to sensitize herpes simplex virus (ASHE et al., 1968, 1969), poliovirus (KELLER, 1968), adenovirus type 2 (WADELL, 1970) and T2 bacteriophage (STEMKE and LENNOX, 1967). Both purified L and H chains of rabbit anti-T1 bacteriophage antibody have been shown to sensitize the virus equally well, whereas the neutralizing activity of the L chain is considerably reduced. The Fd-piece of the anti-bacteriophage antibody sensitized the virus as well (GOODMAN and DONCH, 1965).

Bivalent F(ab)2 fragments obtained by pepsin digestion retain also their sensitizing capacity (WADELL, 1970).

D. Kinetics of the Sensitization Reaction

The results of studies on the kinetics of the sensitization reaction can be summarized as follows: The degree of virus sensitization is proportional to the concentration of antiviral antibodies (ASHE and NOTKINS, 1966; HAHON, 1970; MAJER and LINK, 1970), to the incubation temperature (NOTKINS, 1968) and to the incubation time (ASHE and NOTKINS, 1967; NOTKINS et al., 1968; DUDLEY et al., 1970; MAJER and LINK, 1970). Virus sensitization with neutralizing antisera always precedes the neutralization reaction (ASHE and

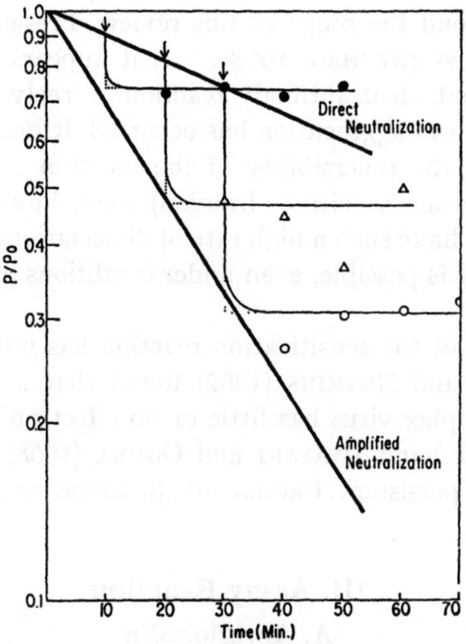

Fig. 1. Enhancement of neutralization (amplified neutralization) of the f2 bacteriophage by antiglobulin added 10 min (●——●), 20 min (△——△), and 30 min (○——○) after initiation of the reaction. The arrows indicate the points at which antiglobulin was added to aliquots of virus-antivirus mixtures (DUDLEY et al., 1970)

NOTKINS, 1967; NOTKINS et al., 1968; DUDLEY et al., 1970; HAHON, 1970; MAJER and LINK, 1970). Presently available experimental data on the kinetics of the sensitization reaction do not permit precise statements about the nature of the reaction. The kinetics of amplified neutralization of an *E. coli* bacteriophage is presented in Fig. 1.

The degree of sensitization of a virus preparation can be measured by the titration of the antiglobulin in the assay reaction. Herpes simplex virus, sensitized with a fourfold higher concentration of anti-virus antibodies, revealed a 256-fold higher titre of the antiglobulin as measured by 50% reduction of infectivity in the assay reaction (DANIELS et al., 1970a).

It was early recognized that sensitized virus is resistant to further neutralization with antiviral antibodies (NOTKINS et al., 1966). Neutralization kinetics showed that, as the degree of sensitization increased, the neutralization rate constant decreased up to 87% of its starting value (Ashe and Notkins, 1967). If further neutralization of sensitized virus was performed with Fab antivirus antibody fragments, the reduction of the neutralization rate constant was more than ten times less (Ashe et al., 1969).

E. Reversibility of the Sensitization Reaction

The question, whether the interaction of virus with neutralizing antibodies is a reversible or irreversible reaction was the subject of various studies in the past, which are beyond the scope of this review. Present knowledge may be summarized as follows (Svehag, 1968): "... it appears highly probable that the initial monovalent virus-antibody reaction is freely reversible. But when multivalent binding and aggregation has occurred, it becomes extremely difficult to demonstrate the reversibility of the reaction ... A small fraction of antibodies (the 'dilution sensitive' fraction) may, however, form complexes with the virus which have such a high rate of dissociation that a demonstration of their dissociability is possible, even under conditions of multivalent binding and aggregation."

The reversibility of the sensitization reaction has not been studied extensively so far. Ashe and Notkins (1967) found that a 1000-fold dilution of sensitized herpes simplex virus has little or no effect on the degree of sensitization. On the other hand Iwasaki and Ogura (1968) found the degree of sensitization of the persistent fraction of Japanese encephalitis virus to be sensitive to dilution.

III. Assay Reaction

A. Antiglobulin

To neutralize sensitized virus, antiglobulin has to react specifically with globulins of the antiviral antiserum. It is usually obtained by heteroimmunization with whole serum (Notkins et al., 1966; Ashe et al., 1966, 1967; Wadell, 1970; Weeke-Lüttmann, 1970) or by its gammaglobulin fractions (Hampar et al., 1968; Notkins et al., 1968; Ashe et al., 1969; Dudley et al., 1970; Hahon, 1970; Majer and Link, 1970; Wallis and Melnick, 1970). Strictly speaking such antiglobulins are not anti-antibodies (see review Gell and Kelus, 1967), although some authors used this term (Dudley et al., 1970; Majer and Link, 1970). Heterologous antiglobulin may cross-react to some extent with sensitized virus, as was the case with anti-human globulin and Venezuelan equine eucephalitis virus sensitized with monkey serum. However anti-chicken, anti-guinea-pig, anti-rabbit, anti-bovine and anti-mouse globulins failed to neutralize significantly the virus sensitized with monkey serum (Hahon, 1970).

The required degree of specificity of antiglobulin was further studied with different rabbit allotypes (Ashe et al., 1968). Herpes simplex virus was sensitized with antisera from a3 b4 and a3 b5 rabbits. Virus sensitized with antibodies from one rabbit allotype was inactivated only by the antiglobulin to this allotype, but not by the antiglobulin to the other. The Ab allotypic determinants are localized on the K light chain of the rabbit IgG molecule (see review KRAUSE, 1970), thus indicating the locus of interaction with the antiglobulin. On the other hand the allotypic determinant of BALB/C mice was found to be localized on the Fc fragment of the IgG molecule (DRAY et al., 1965; LIEBERMAN et al., 1965; POTTER et al., 1966). Japanese encephalitis virus sensitized with BALB/C mouse serum was inactivated by the allotype antiglobulin showing that in this case the antibody was determined against the Fc fragment of the antiviral antibody molecule (IWASAKI and OGURA, 1968).

The localization of the binding site for the antiglobulin in the assay reaction was also studied with antifragment antibodies. Anti-Fab immunoglobulin could neutralize sensitized virus. Anti-Fc globulin neutralized the sensitized virus ten times less effectively than antiserum to Fab fragments. The mixture of both anti-Fab and anti-Fc antibodies neutralized almost as much virus as anti-IgG globulin (HAHON, 1970). Anti-human μ chain and anti-human α chain sera as well as the rheumatoid factor failed to neutralize sensitized herpes simplex virus (DANIELS et al., 1970). The rheumatoid factor (an IgM antiglobulin) had in fact attached to the complex as was shown by neutralization of the virus infectivity after addition of anti-IgM globulin (ASHE et al., 1971).

To study the mechanism of the assay reaction, antiglobulin fragments obtained with papain digestion and separated by DEAE-cellulose chromatography were used (NOTKINS et al., 1968). The Fab fragments of goat anti-mouse globulin partially neutralized lactic dehydrogenase virus sensitized with whole mouse serum, whereas Fc fragment had no effect. Univalent Fab fragments of sheep anti-rabbit gammaglobulin were able to neutralize sensitized herpes simplex virus (ASHE et al., 1968). Here again the fragments were less effective than the intact antiglobulin.

The antiglobulin can be titrated in the assay reaction. Its titre will depend on the amount of anti-viral antibodies used in the sensitization reaction, the concentration of sensitized virus and the amount of unbound antivirus serum (DANIELS et al., 1970a). The increase of degree of virus sensitization will cause an apparent increase in the antiglobulin titre, whereas the increase in the titre of sensitized virus will cause an apparent decrease in the antiglobulin titre.

B. Kinetics of the Assay Reaction

To study the kinetics of neutralization of sensitized virus by antiglobulin, the preceding sensitization reaction has to achieve its equilibrium to exclude additional neutralization by antiviral antibodies. The rate of neutralization by antiglobulin is proportional to its concentration (IWASAKI and OGURA, 1968). The effective antiglobulin concentration will depend on the quantity of added antiglobulin and on the concentration of free globulin in the sensitized virus preparation, if free globulin has not been removed by differential centrifugation. Incubation temperature seems not to be a significant factor (HAHON, 1970). Sensitized lactic dehydrogenase virus was nearly completely neutralized after 1 minute of incubation with antiglobulin at 4°C (NOTKINS et al., 1968). High reaction rate is a striking property of the assay reaction. Sensitized herpes simplex virus was substantially neutralized within 15 seconds at incubation temperature of 37°C (ASHE and NOTKINS, 1966).

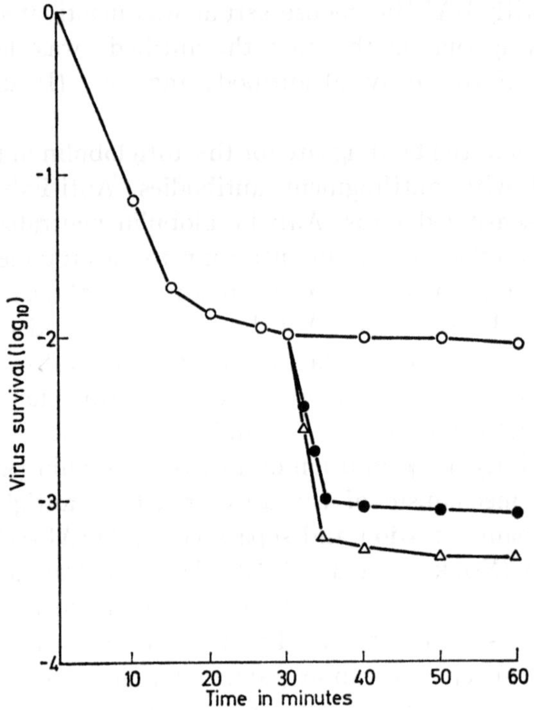

Fig. 2. Kinetics of neutralization of the persistent fraction of JEV by heterotype (●——●) or allotype (△——△) antiglobulin (IWASAKI and OGURA, 1968)

The kinetics of neutralization of sensitized Japanese encephalitis virus by heterotype and allotype antiglobulin is shown in Fig. 2. Both antiglobulin preparations neutralized the virus by first-order reaction (IWASAKI and OGURA, 1968).

IV. Mechanism of Virus Sensitization

The mechanism of virus sensitization will be discussed in connection with the mechanism of neutralization by antiglobulin. Both reactions are based on the antigen-antibody binding, which mechanism will, however, not be considered here. Its effects on the infectivity of single virus particles will be analysed. The first problem is the fact that virus suspensions and not single particles are usually under study. Virus suspensions may contain natural virus aggregates and they may form immuno-aggregates after interaction with antibodies. In both cases the fate of single viria can hardly be estimated. The use of filtered mono-dispersed and highly diluted virus suspensions minimize these complications. Further strong evidence that the phenomena to be discussed relate to single viria are filtration experiments with virus-antibody complexes (ASHE et al., 1969; WALLIS and MELNICK, 1970) and the use of univalent antibody fragments in both sensitization and assay reaction (ASHE et al., 1968; NOTKINS et al., 1968). Furthermore, certain test systems for virus infectivity as the end point dilution test for measuring antibodies will discriminate between loss of infectivity and immunoaggregation (WADELL, 1970).

The understanding of virus sensitization is made more difficult by the fact that the mechanism of virus neutralization, which has been studied for the last few decades, is not yet completely understood. The neutralization seems to be best defined as "Antibody-mediated inhibition of the expression of the viral genome as indicated by absence of viral replication" (WADELL, 1970). An alternative mechanism is the irreversible damage of viria by antibodies. In special cases, like virolysis, the latter mechanism may in fact operate (BERRY and ALMEIDA, 1968). For the neutralization of sensitized virus with antiglobulin there are no experimental data in favour of one or the other explanation. However the first definition is generally applied to the neutralization of sensitized virus. This does not mean that virus sensitization is necessarily a stage of the neutralization reaction. Nevertheless regarding the timing of the processes, virus sensitization precedes the neutralization reaction. It should be stressed here that in most studies heat inactivated antibody preparations have been used indicating that thermolabile serum factors are not required either for the sensitization or for the assay reaction.

Two hypotheses on the mechanism of virus sensitization are summarized as follows: According to one hypothesis, virus sensitization is a stage of the neutralization process which in some instance may not proceed beyond the sensitization reaction, thus leading to the persistent fraction. The absence of virus neutralization is explained by dissociation of virus-antibody complexes in the reaction mixture or at the level of the host cell. The inactivating effect of antiglobulin may be explained by stabilization of virus-antibody complexes. Non-avid antibodies could be involved in the formation of the persistent fraction, whereas virus sensitization which precedes the neutralization reaction could be understood as a fairly loose-binding between virus and antibodies of average avidity at an early stage of virus-antibody interaction. The existence of antibodies of low avidity and the dissociation of virus-antibody complexes cannot be questioned (see review SVEHAG, 1968), however, conclusive evidence for the validity of the outlined hypothesis is still missing. Nevertheless it may be applied to some cases, like the non-neutralizable fraction of Japanese encephalitis virus (IWASAKI and OGURA, 1968). However contradictory results have been reported for herpes simplex virus (ASHE and NOTKINS, 1967). The understanding of the proposed stabilization of virus-antibody complexes by antiglobulin is not satisfactory. The antiglobulin might form bridges between two antivirus antibody molecules bound to the same virus particle giving the same effect as the stable two-site attachment of antiviral antibodies (LAFFERTY, 1963a; IWASAKI and OGURA, 1968). This will help us to understand the neutralization by bivalent antiglobulin of virus-sensitized with univalent antivirus fragments, but it is hardly applicable to the neutralization by univalent antiglobulin.

According to the other hypothesis, virus sensitization is a distinct process of antibody attachment to "non-critical" sites of virus surface, whereas virus neutralization would result from antibodies bound to "critical" sites (NOTKINS et al., 1966). In this context, critical sites are areas on the virus surface involved in the initiation of the infection. Critical and non-critical sites may bear the same or different antigenic determinants. The inactivating effect of antiglobulin would result from blocking the neighbouring critical sites after its attachment to antiviral antibodies bound to non-critical ones. Fewer antiglobulin molecules seem to be needed to neutralize a virion when large amounts of antivirus antibodies are attached to it. As expected, univalent fragments of the antiglobulin, being of smaller size, are thereby less effective (ASHE et al., 1968; NOTKINS et al., 1968). With this hypothesis one could try to explain also other experimental findings. Antibodies bound to non-critical sites would hinder the attachment of antibodies to critical sites, thus leading to a reduced neutralization rate constant (ASHE and NOTKINS, 1967) and finally to the well-known persistent fraction. Sensitization prior to neutralization could be explained by proposing the existence of more non-critical sites than critical ones (RAPPAPORT, 1970). Finally, virus sensitization by non-neutralizing antibodies, which are directed merely against non-critical sites, is easily understandable in the light of this hypothesis.

V. Prospects of Virus Sensitization

Virus sensitization represents a serological test system which offers some practical advantages over the conventional techniques. The estimation of sensitized virus by antiglobulin is a very sensitive method for detecting antiviral antibodies. This applies to neutralizing antisera to different viruses (Fig. 3) as well as to antibodies which themselves have no measurable neutralizing effect (WADELL, 1970; MAJER and LINK, 1971).

These properties make the test suitable for revealing immunologic cross-reactions between viruses (KJELLÉN and PEREIRA, 1968), and for detecting contamination of antibody preparations which would not be detected by other methods (ASHE et al., 1968). The test also might be suitable for early diagnosis of viral diseases (UNGAR, 1971). Furthermore, some problems imposed by incomplete neutralization may be overcome by the use of antiglobulin. When testing live smallpox vaccines for adventitious viruses, the non-neutralizable fraction may interfere with the test (ESPMARK, 1966), whilst an extensive specific inactivation of the sensitized vaccinia virus may be achieved by antiglobulin (MAJER and LINK, 1970). The "break-through" phenomenon which may complicate antibody titration in the end point dilution test can also be eliminated by applying the indirect neutralization assay (Majer and Link, 1970). Sensitized virus was also used to titrate various antiglobulin preparations (Daniels et al., 1970a). Regarding its sensitivity the test was superior to the conventional passive haemagglutination technique.

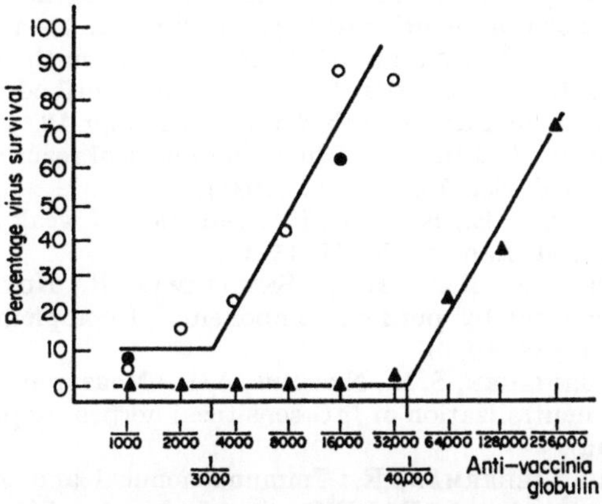

Fig. 3. Titration of an anti-vaccinia globulin by the direct (○——○) and by the indirect (▲——▲) neutralization assay. Using the "inversion point" (Link and Majer, 1970) as the endpoint of the titration, the immunoglobulin had a titre of 1/40000 in the indirect test compared with a titre of only 1/3000 in the direct test (Majer and Link, 1970)

Since virus sensitization also occurs in vivo (Notkins et al., 1966; Porter and Larsen, 1967; Oldstone and Dixon, 1971), it probably plays a role in the pathogenesis of certain diseases. It was recently shown that virus-antibody complexes cause renal lesions in immune-complex diseases (Hirsch et al., 1969; Oldstone and Dixon, 1969; Porter et al., 1969). Further virus-antibody complexes might induce the production of anti-antibodies, which could influence the severity of illness. In a preliminary experiment, antiglobulin was able to protect mice against a challenge of sensitized influenza virus (Zalan et al., 1971).

References

Adler, H. E., DaMassa, A. J.: Enhancement of mycoplasma agglutination titers by use of antiglobulin. Proc. Soc. exp. Biol. (N.Y.) **116**, 608–610 (1964).

Almeida, J. D., Cinder, B., Howatson, A. F.: The structure of antigen-antibody complexes. A study by electron microscopy. J. exp. Med. **118**, 327–340 (1963).

Andrewes, C.: Viruses of vertebrates. London: Baillère, Tindall and Cox 1964.

Andrewes, C. H., Elford, W. J.: Observations of antiphage sera. I: "The percentage law." Brit. J. exp. Path. **14**, 367–376 (1933).

Andrewes, C. H., Elford, W. J.: Observations on antiphage sera. II: Properties of incompletely neutralized phage. Brit. J. exp. Path. **14**, 376–383 (1933a).

Ashe, W. K., Daniels, C. A., Scott, G. S., Notkins, A. L.: Interaction of rheumatoid factor with infectious herpes simplex virus-antibody complexes. Science (Wash.) **172**, 176–177 (1971).

Ashe, W. K., Mage, M., Mage, R., Notkins, A. L.: Neutralization and sensitization of herpes simplex virus with antibody fragments from rabbits of different allotypes. J. Immunol. **101**, 500–504 (1968).

Ashe, W. K., Mage, M., Notkins, A. L.: Kinetics of neutralization of sensitized herpes simplex virus with antibody fragments. Virology **37**, 290–293 (1969).

Ashe, W. K., Notkins, L.: Neutralization of an infectious herpes simplex virus-antibody complex by anti-γ-globulin. Proc. nat. Acad. Sci. (Wash.) **56**, 447–451 (1966).

Ashe, W. K., Notkins, A. L.: Kinetics of sensitization of herpes simplex virus and its relationship to the reduction in the neutralization constant. Virology **33**, 613–617 (1967).

Berry, D. M., Almeida, J. D.: The morphological and biological effects of various antisera on avian infectious bronchitis virus. J. gen. Virol. **3**, 97–102 (1968).

Bradish, C. J., Farley, J. O., Ferrier, H. E. N.: Studies on the nature of the neutralization reaction and the competition for neutralizing antibody between components of the virus system of foot-and-mouth disease. Virology **18**, 378–400 (1962).

Burnet, F. M., Keogh, E. V., Lush, D.: The immunological reactions of filtrable viruses. Aust. J. exp. Biol. med. Sci. **15**, 227–247 (1937).

Coombs, R. R., Mourant, A. E., Race, R. R.: Detection of weak and "incomplete" Rh agglutinins: A new test. Lancet **1945 II**, 15–16.

Daniels, Ch. A., Borsos, T., Rapp, H. J., Snyderman, R., Notkins, A. L.: Neutralization of sensitized virus by purified components of complement. Proc. nat. Acad. Sci. (Wash.) **65**, 528–535 (1970).

Daniels, Ch. A., Mergenhagen, S. E., Notkins, A. L.: Measurement of antibody against human IgG by the neutralization of IgG-sensitized herpes simplex virus. J. Immunol. **104**, 470–474 (1970a).

Dray, S., Potter, M., Lieberman, R.: Immunochemical and genetic studies of two distinct G-immunoglobulins in BALB/C mice. J. Immunol. **95**, 823–833 (1965).

Dudley, M. A., Henkens, R. W., Rowlands, D. T.: Kinetics of neutralization of bacteriophage f2 by rabbit G-antibodies. Proc. nat. Acad. Sci. (Wash.) **65**, 88–95 (1970).

Dulbecco, R., Vogt, M., Stickland, A.: A study of the basic aspects of neutralization with two animal viruses, western equine encephalitis virus and poliomyelitis virus. Virology **2**, 162–205 (1956).

Espmark, J. Å.: Smallpox vaccine: attributes of egg and calf lymph vaccines. First International Conference on Vaccines against Viral and Rickettsial Diseases of Man, Washington, 1966.

Fazekas de St. Groth, S.: The neutralization of viruses. Advanc. Virus Res. **9**, 1–116 (1962).

Friedman, M.: The effect of the host of the properties and sensitivity of a bacterial virus. J. Bact. **68**, 274–278 (1954).

Gell, P. H. G., Kelus, A. S.: Anti-antibodies. Advanc. Immunol. **6**, 461–478 (1967).

Goodman, J. W., Donch, J. J.: Phage-neutralizing activity in light polypeptide chains of rabbit antibody. Immunochemistry **2**, 351–357 (1965).

Gudnadóttir, M., Pálsson, P. A.: Host-virus interaction in Visna infected sheep. J. Immunol. **95**, 1116–1120 (1965).

Hahon, N.: Neutralization of residual infectivity of Venezuelan equine encephalomyelitis virus by anti-gamma globulin. J. gen. Virol. **6**, 361–372 (1970).

Hampar, B.: Persistent cyclic herpes simplex virus infection in vitro. II. Localization of virus, degree of cell destruction, and mechanism of virus transmission. J. Bact. **91**, 1959–1964 (1966).

Hampar, B., Notkins, A. L., Mage, M., Keehn, M.: Heterogeneity in the properties of 7S and 19S rabbit neutralizing antibodies to herpes simplex virus. J. Immunol. **100**, 586–593 (1968).

Hashimoto, L., Prince, A. M.: Kinetic studies on the neutralization reaction between Japanese encephalitis virus and antiserum. Virology **19**, 261–272 (1963).

Hirsch, M. S., Allison, A. C., Harvey, J. J.: Immune complexes in mice infected neonatally with Molone leukaemogenic and murine sarcoma viruses. Nature (Lond.) **223**, 739–740 (1969).

Itoh, H., Melnick, J. L.: The infection of chimpanzees with echoviruses. J. exp. Med. **106**, 677–688 (1957).

Iwasaki, T., Ogura, R.: Studies on neutralization of Japanese encephalitis virus (JEV) I. Further neutralization of the resistant virus fraction by an interaction between anti-virus IgG antibody and IgG heterotype or allotype antibody. Virology **34**, 141–148 (1968).

KELLER, R.: Studies on the mechanism of the enzymic reactivation of antibody-neutralized poliovirus. J. Immunol. **100**, 1071–1079 (1968).
KELLER, R., DWYER, J. E.: Neutralization of poliovirus by IgA coproantibodies. J. Immunol. **101**, 192–202 (1968).
KJELLÉN, L.: Density gradient centrifugation of adenovirus-antibody complexes. Arch. ges. Virusforsch. **17**, 398–408 (1965).
KJELLÉN, L., PEREIRA, H.H.: Role of adenovirus antigens in the induction of virus neutralizing antibody. J. gen. Virol. **2**, 177–185 (1968).
KJELLÉN, L., SCHLESINGER, R. W.: Influence of host cell on residual infectivity of neutralized vesicular stomatitis virus. Virology **2**, 236–251 (1959).
KRAUSE, R. M.: The search for antibodies with molecular uniformity. Advanc. Immunol. **12**, 1–56 (1970).
LAFFERTY, K. J.: The interaction between virus and antibody. I. Kinetic studies. Virology **21**, 61–75 (1963).
LAFFERTY, K. J.: Interaction between virus and antibody. II. Mechanism of the reaction. Virology **21**, 76–90 (1963a).
LEDINKO, N., MELNICK, J. L.: Poliomyelitis virus in tissue culture. V. Reaction of virus and antibody; variables of the quantitative neutralization test. Amer. J. Hyg. **58**, 223–247 (1953).
LIEBERMAN, R., DREY, S., POTTER, M.: Linkage in control of allotypic specificities on two different gamma G-immunoglobulins. Science **148**, 640–642 (1965).
LINK, F., MAJER, M.: Contribution to the standardisation of the biological control of anti-smallpox immunoglobulins (antisera). Proceedings of the 11th Internat. Congr. for Microbiological Standardisation, Milan 1968. Vol. 4 of the Progress in immunobiological standardisation, p. 141–144. Basel–München–New York: Karger 1970.
MAJER, M., LINK, F.: Untersuchungen über die nicht neutralisierbare Fraktion des Vaccinia-Virus. Annual Meeting of the Swiss Society for Immunology, Geneva, 1968 (unpublished).
MAJER, M., LINK, F.: Studies on the non-neutralizable fraction of vaccinia virus. Clin. exp. Immunol. **7**, 283–291 (1970).
MAJER, M., LINK, F.: Sensitization of influenza virus A2/Singapore by antiglobulin. J. gen. Virol. **13**, 355–356 (1971).
MAJER, M., LINK, F.: Studies on a recombinant influenza A virus by indirect neutralization. Z. Immun.-Forsch. (1972 in press).
MANDEL, B.: Studies on the interaction of poliomyelitis virus, antibody, and host cell in a tissue culture system. Virology **6**, 424–447 (1958).
MARRACK, J. R.: The chemistry of antigens and antibodies. Report No 230, Med. Research Council, p. 116–118. London: H. M. Stationary Office 1938.
MELNICK, J. L.: Application of tissue culture methods to epidemiological studies of poliomyelitis. Amer. J. publ. Hlth **44**, 571–580 (1954).
NORRBY, E., MARUSYK, H., HAMMARSKJÖLD, M. L.: The relationship between the soluble antigens and the virion of adenovirus type 3. V. Identification of antigen specificities available at the surface of virions. Virology **38**, 477–482 (1969).
NOTKINS, A.L.: Infectious virus-antibody complexes: Interaction with anti-immuno-globulins, complement, and rheumatoid factor. J. exp. Med. **134** (No 3, pt. 2) 41s–51s (1971).
NOTKINS, A. L., MAGE, M., ASHE, W. K., MAHAR, S.: Neutralization of sensitized lactic dehydrogenase virus by anti-gamma-globulin. J. Immunol. **100**, 314–320 (1968).
NOTKINS, A. L., MAHAR, S., SCHEELE, CH., GOFFMAN, J.: Infectious virus-antibody complex in the blood of chronically infected mice. J. exp. Med. **124**, 81–97 (1966).
NOTKINS, A. L., ROSENTHAL, I., JOHNSON, B.: Rate-zonal centrifugation of herpes simplex virus-antibody complexes. Virology **43**, 321–325 (1971).
OLDSTONE, M. A. B., DIXON, F. J.: Pathogenesis of chronic disease associated with persistent lymphocytic choriomeningitis viral infection. I. Relationship of antibody production to disease in neonatally infected mice. J. exp. Med. **129**, 483–503 (1969).
OLDSTONE, M. B. A., DIXON, F. J.: Complexes in viral infections. J. exp. Med. **134** (No 3, pt. 2) 32–40 (1971).
PAULING, L., PRESSMAN, D., GROSSBERG, A. L.: The serological properties of simple substances. VII. J. Amer. chem. Soc. **66**, 784–792 (1944).
PHILIPSON, L.: Interaction of poliovirus and immunoglobulins. II. Basic aspects of virus-antibody interaction. Virology **28**, 35–46 (1966).

PORTER, D. D., LARSEN, A. E.: Aleutian disease of mink: Infectious virus-antibody complexes in the serum. Proc. Soc. exp. Biol. (N.Y.) **126**, 680–682 (1967).

PORTER, D. D., LARSEN, A. E., PORTER, H. G.: The pathogenesis of Aleutian disease of mink. I. In vivo viral replication and the host antibody response to viral antigen. J. exp. Med. **130**, 575–589 (1969).

PORTER, R. R.: The hydrolysis of rabbit gamma-globulin and antibodies with crystalline papain. Biochem. J. **73**, 119–126 (1959).

POTTER, M., LIEBERMAN, R., DRAY, S.: Isoantibodies specific for myeloma gamma-G and gamma-H immunoglobulins of BALB/C mice. J. molec. Biol. **16**, 334–346 (1966).

RAPPAPORT, I.: The antibody-antigen reaction. A hypothesis to account for the presence of uncombined antigenic sites in the presence of excess antibody. J. Immunol. **78**, 246–255 (1957).

RAPPAPORT, I.: An analysis of the inactivation of MS2 bacteriophage with antiserum. J. gen. Virol. **6**, 25–32 (1970).

RUBIN, H., FRANKLIN, R. M.: On the mechanism of Newcastle disease virus neutralization by immune serum. Virology **3**, 84–95 (1957).

SCHILD, G. C., MCCAHON, D., KENDAL, A. P.: Immunological and structural studies on a recombinant influenza A virus. In: The biology of large RNA viruses, BARRY, R. D. and B. W. J. MAHY, eds., p. 638–654. London-New York: Academic Press 1970.

SMITH, K. O., TRONSDALE, G., GEHLE, W.: Adenovirus-antibody agglutination reaction J. Immunol. **95**, 810–817 (1965).

STEMKE, G. W., LENNOX, E. S.: Bacteriophage neutralizin activity of fragments derived from rabbit immunoglobulin by papain digestion. J. Immunol. **98**, 94–101 (1967).

STERNBERG, G. M. (1892): Cited by BOULTER, E. A., The titration of vaccinal neutralizing antibody on chorio-allantoic membranes. J. Hyg. (Lond.) **55**, 502–512 (1957).

SVEHAG, S. E.: Diversity of antibodies formed against viruses. In: Antibodies to biologically active molecules, CINADER, B., ed., p. 301–348. New York: Pergamon Press 1966.

SVEHAG, S. E.: Formation and dissociation of virus-antibody complexes with special reference to the neutralization process. Progr. med. Virol. **10**, 1–63 (1968).

TOOLAN, H. W.: H-1 virus viraemia in adult hamster. Proc. Soc. exp. Biol. (N.Y.) **119**, 715–717 (1965).

UNGAR, J.: Diagnosis of viral infections. 12th Pugwash Symposium on rapid detection and identification of microbiological agents. Geneva, 1971.

WADELL, G.: Aspects of the neutralization of adenoviruses. J. gen. Microbiol. **57**, XX (1969).

WADELL, G.: Structural and biological properties of capsid components of human adenoviruses. Doctoral Thesis, Karolinska Institute, Stockholm. 1970.

WALLIS, C., MELNICK, J. L.: Virus aggregation as the cause of the nonneutralizable persistent fraction. J. Virol. **3**, 478–488 (1967).

WALLIS, C., MELNICK, J. L.: Herpesvirus neutralization: Induction of the persistent fraction by insufficient antibody. Virology **42**, 128–137 (1970).

WEBSTER, R. G., LAVER, W. G.: Preparation and properties of antibody directed specifically against the neuraminidase of influenza virus. J. Immunol. **99**, 49–55 (1967).

WEEKE-LÜTTMANN, M.: Steigerung der Neutralisation von Vaccinia-Virus durch Antiglobulin. Arch. ges. Virusforsch. **31**, 28–32 (1970).

WELLER, T. H., COONS, A. H.: Fluorescent antibody studies with agents of varicella and herpes zoster propagated in vitro. Proc. Soc. exp. Biol. (N.Y.) **86**, 789–794 (1954).

WESTAWAY, E. G.: The neutralization of arboviruses. I. Neutralization in homologous virus-serum mixtures with two group B arboviruses. Virology **26**, 517–527 (1965).

WESTAWAY, E. G.: Antibody responses in rabbits to the group B arbovirus Kunjin: Serologic activity of the fractionated immuno-globulins in homologous and heterologous reactions. J. Immunol. **100**, 569–580 (1968).

YOSHINO, K., TANIGUCHI, S.: Studies on the neutralization of herpes simplex virus. I. Appearance of neutralizing antibodies having different grades of complement requirement. Virology **26**, 44–53 (1965).

YOSHINO, K., TANIGUCHI, S.: Studies on the neutralization of herpes simplex virus. III. Mechanisms of the antibody-potentiating action of complement. Virology **26**, 61–72 (1965a).

ZALAN, E., BORMAN, E., LABZOFFSKY, N. A.: Infectious influenza virus-antibody complex. Arch. ges. Virusforsch. **34**, 209–213 (1971).

Antigenic Determinants of Proteins of Defined Sequences

E. Benjamini[1], D. Michaeli[2], and J. D. Young[3]

Contents

I. Introduction

Inspired by the classical work of Landsteiner (1945), many years of effort have been directed towards the better understanding, on a molecular level, of various parameters of the immune response. To facilitate such studies, various model antigens have been devised and successfully used. These include hapten-carrier conjugates, synthetic polypeptides, carbohydrates, and protein

1 Department of Medical Microbiology, School of Medicine, University of California, Davis.
2 Department of Biochemistry and Surgery, University of California, San Francisco; Recipient of U.S.P.H.S. Res. Career Devel. Award No. 5-K4-AM-50, 205.
3 Department of Bacteriology and Immunology, University of California, Berkeley.

antigens. The studies have been the subject of numerous reviews (Karush, 1962; Kaminski, 1965; Singer, 1965; Sela, 1966, 1967, 1969; Crumpton, 1967b; Kabat, 1968; Pressman and Grossberg, 1968; Goodman, 1969).

Polypeptides of known amino acid sequence and/or structure afford powerful tools for investigating the chemical aspects underlying the immune response. The 1960's witnessed a proliferation of protein characterization; it is therefore not surprising that many such characterized proteins and peptides have been utilized as models in immunochemical studies.

In this article we will focus our attention on immunochemical findings obtained from studies with several proteins and peptides of known amino acid sequence and/or structure, and attempt to integrate the diverse findings in order to elucidate some of the principles underlying antigen-antibody interaction and immunologic specificity.

This article is divided into two main parts: the first deals with a summary of findings on each of several representative proteins; in the second part we attempt to integrate these findings and to formulate some general principles on the relationship between structure and antigenic specificity. Particular emphasis will be placed on data dealing with the interaction of antigenic sequences with antibodies, and on the relationship between structure and activity. In this latter section, we have taken the liberty to reference only supplementary information which was not previously referenced.

Since only a few representative proteins are dealt with in the present chapter, we extend our apologies to those workers whose material was not covered. Also if there are any important omissions, they were strictly unintentional.

II. Antigens

Adrenocorticotropic Hormone (ACTH)

10

SER–TYR–SER–MET–GLU–HIS–PHE–ARG–TRP–GLY–LYS–PRO–

20

VAL–GLY–LYS–LYS–ARG–ARG–PRO–VAL–LYS–VAL–TYR–PRO–

30

ASP–ALA–GLY–GLU–ASP–GLN–SER–ALA–GLU–ALA–PHE–PRO–

39

LEU–GLU–PHE (Human ACTH, according to Lee et al., 1961).

Adrenocorticotropic hormone is a straight chain peptide of 39 amino acids. The sequences of ACTH derived from several species are known (Dayhoff and Eck, 1969). Of the 39 amino acids, the N-terminal sequence 1–24 is common to all known ACTH and is responsible for the hormonal activity of the molecule. Species variation is found in residues 25–33 of the hormone; the C-terminal hexapeptide is again common to all the species studies.

Probably due to its relatively low molecular weight, the hormone is a poor immunogen. Consequently, antibodies to either the natural or the synthetic hormone are commonly elicited in response to immunization with the hormone, or fragments thereof, conjugated to protein carriers (FISCHER et al., 1965; FELBER et al., 1966; FELBER and MICHELI, 1967; SALVIN and LIAUW, 1967; GELZER, 1968; REICHLIN et al., 1968b). The C-terminal part of the hormone (sequences 25–39) possessed the highest inhibitory effect on the binding between ACTH and homologous antibodies (FLEISCHER et al., 1965; IMURA et al., 1965; FLEISCHER et al., 1966; GELZER, 1968). Other workers have demonstrated antibodies which were capable of neutralizing the biological activity of the hormone, and showed that these antibodies were directed to sequences 1–24 (FELBER et al., 1966; FELBER and MICHELI, 1967). Moreover, the free 1–24 peptide was found to be immunogenic in guinea pigs (AXELROD et al., 1963; SALVIN and LIAUW, 1967). In addition, it was demonstrated that whereas the peptide 1–23 of ACTH was not immunogenic, the same peptide, but N-acetylated and formylated at the ε amino groups of lysines was void of hormonal activity but was immunogenic (AXELROD et al., 1963).

The antigenic specificity of peptide 1–24 was investigated by GELZER, who utilized rabbit antisera to immunigens consisting of peptide 1–24, or various derivatives thereof, coupled to rabbit serum albumin (GELZER, 1968). Compared to the inhibitory activity of the 1–24 peptide, the C-terminal 11–24 and 17–24 portions were stronger inhibitors (confirming a similar observation by IMURA et al., 1965), whereas the C-terminal 20–24 peptide possessed very little activity. The conclusion was drawn that the combining site of the antibody was complementary to the octapeptide or the tetradecapeptide representing sequences 17–24 and 11–24 respectively. However, the actual size of the antigenic determinant may be smaller than the above peptides since the substitution of the two arginines at positions 17 and 18 with ornithine did not drastically affect antigenicity. This suggests that the hexapeptide consisting of sequence 19–24 may constitute the determinant.

Angiotensin

ASP–ARG–VAL–TYR–ILE–HIS–PRO–PHE–HIS–LEU (Horse angiotensin, according to SKEGGS et al., 1957).

Angiotensin (also known as hypertensin, angiotonin) is a tissuc hormone or kinin, found in the kidney. The most potent pressor substance known, it is formed from angiotensinogen (an inactive precursor molecule) by the action of renin, which releases an active decapeptide (angiotensin I) having the sequence shown above. This may be further converted to angiotensin II (also an active peptide) following the enzymatic removal of the C-terminal dipeptide.

Probably because of its small size and susceptibility to enzymatic degradation, free angiotensin is usually not immunogenic. Thus, angiotensin conjugated to carriers is normally used for antibody production (DEODHAR, 1960; GOODFRIEND et al., 1964; HABER et al., 1965; GOODFRIEND et al., 1966).

However, immunization with free angiotensin II amide (Asp-β amide) physically adsorbed onto microparticles of carbon did result in the production of antibodies to free angiotensin. These antibodies could also effectively neutralize hormonal activity (Boyd and Peart, 1968).

The immunological activity of angiotensin and its analogues is usually assessed by their ability to inhibit the reaction either between the polyvalent angiotensin-carrier conjugate and antibodies, or between isotopically labeled free angiotensin and antibodies. Recently angiotensin II conjugated to erythrocytes was used for analysis of antibodies to angiotensin by passive hemagglutination (Dietrich and Frischknecht, 1968). In addition, the inactivation, by antiangiotensin, of bacteriophage T_4 to which angiotensin was conjugated, and the inhibition of this inactivation by angiotensin afford a sensitive assay for angiotensin and for antibodies to angiotensin (Hurwitz et al., 1970). Assays of the immunological activity of several analogues of Val 5 angiotensin, revealed that the phenolic hydroxyl at position 4, and the two N-terminal residues were important for antigenicity. Moreover, disruption of the chain conformation by substituting the L-with the D-isomer at position 4 resulted in greatly reduced antigenic activity (Goodfriend et al., 1966). On the other hand, other studies have indicated that the heptapeptide 2–8 or the hexapeptide 3–8 were responsible for antigenic activity and that the residue at position 4 was not critical (Dietrich, 1967). It was further concluded by Dietrich that the C-terminal phenylalanine was of utmost importance for immunological specificity. This latter finding is corroborated by work of D'Auriac et al. (1969) who showed that angiotensin I did not react with antibodies produced against Val 5 angiotensin II. Also, Catt and Coghlan (1967) showed that antibodies to synthetic angiotensin II reacted only slightly with angiotensin I. The importance of the C-terminal portion of the molecule to immunological activity was also indicated by the work with angiotensin analogues and fragments (Haber et al., 1967). The discrepancies between the various conclusions may be accounted for by differences in the techniques used, and are undoubtedly due to genetic differences in the response of the animals used. Indeed, it was shown that different rabbits, immunized with Asn 1–Val 5 angiotensin II conjugate, produced antibodies with different affinities to Asn 1–Val 5 angiotensin, to Asp 1–Val 5 angiotensin, to Asn 1–Ile 5 angiotensin, and to heptapeptide 2–8 (Hollemans et al., 1968).

The question whether the structural requirements for antigenicity parallel those for the hormonal activity is still somewhat unclear. Although the hormonal activity can be neutralized by antibodies (Deodhar, 1960; Hedwall, 1968; Oken and Biber, 1968) there is evidence that the structural requirements for binding with antibodies differ from those necessary for binding with biological receptors of the hormone; the hexapeptide 3–8, while possessing immunological activity, has only 2% of the pressor activity and the monoiodo and diiodo Tyr 4 which possess immunological activity have only 25% and 1–4% of the pressor activity respectively (Dietrich, 1967).

Poly γ-D-Glutamyl Peptide of the Capsule of *Bacillus anthracis*

$$(-\overset{\overset{O}{\|}}{C}-NH-\overset{\overset{COOH}{|}}{CH}-CH_2-CH_2-\overset{\overset{O}{\|}}{C}-NH-\overset{\overset{COOH}{|}}{CH}-CH_2-CH_2-\overset{\overset{O}{\|}}{C}-NH-)_n$$

The capsular polypeptide from *Bacillus anthracis* consists of D-glutamic acids linked by γ peptide bonds. Immunization with *B. anthracis* gives rise to antibodies with specificity to the γ-D-glutamic acid polypeptide. Although the capsular polypeptide is of molecular weight of 33000 it is non-immunogenic, but is capable of acting as a hapten. Thus, when the peptide is adsorbed onto methylated bovine serum albumin, antibodies with specificity to the poly γ-D-glutamyl peptide are produced (Summarized by Goodman, 1969). Experiments by Roelants and Goodman showed that while the pure poly γ-D-glutamyl peptide is non-immunogenic it may induce tolerance in animals subsequently challenged with the immunogenic form of the polypeptide (Roelants and Goodman, 1970).

The poly γ-D-glutamyl peptide was used by Goodman and his associates to study the relationship between structure of the polypeptide and its reaction with rabbit antiserum. Short polymers of γ-D-glutamic acid were prepared synthetically (Nitecki and Goodman, 1966) and by partial acid hydrolysis of the capsular polypeptide of *B. anthracis* (Goodman et al., 1968). The immunological activity of the peptides was ascertained by their capacity to inhibit the quantitative precipitin reaction between rabbit antisera specific to the capsular polypeptide and the homologous antigen. With 2 sera, increments in inhibiting efficiency were obtained up to a penta-γ-D-Glu peptide, which inhibited to the same degree as the hexapeptide; a third antiserum distinguished between the pentapeptide and the hexapeptide. The results suggested that the optimal size of the antigen involved in binding was equivalent to a hexapeptide, indicating a size of approximately $36 \times 12 \times 7$ Å (the size of hexaglutamic acid). In the same study it was found that mixtures of larger peptides, obtained from the partial acid hydrolysis of the capsular polypeptide, were markedly better inhibitors than the hexa-D-glutamic acid. Several explanations were proposed for this latter finding: a) Antibodies may have been directed against a conformation of the capsular polypeptide which is also possessed by a large poly γ-D-glutamyl peptide but not by a hexapeptide. This is supported by findings that there exists a direct correlation between peptide length and acquisition of secondary structure in solution (Goodman, 1969). b) An alternative explanation for the enhanced activity of larger peptides over that of the hexapeptide may be based on the multiplicity of hexapeptide units in the polypeptide chain; such a multivalent antigen would be expected to be a better inhibitor than the univalent hexa-D-gluamyl hapten.

Bradykinin

ARG–PRO–PRO–GLY–PHE–SER–PRO–PHE–ARG (Bovine, according to Elliott et al., 1960).

Antibodies to bradykinin, a nonapeptide of the sequence shown above, have been obtained by immunization of rabbits with a conjugate consisting of bradykinin coupled to poly-L-lysine through the N-terminal portion of the peptide (SPRAGG et al., 1966).

The relationship between peptide structure and binding with antibodies was evaluated from experiments in which various analogues of the peptide were used for the inhibition of binding between the native peptide and antibodies. These experiments revealed some interesting phenomena regarding antigenic specificity and binding (HABER et al., 1967; SPRAGG et al., 1967; SPRAGG et al., 1968). The length of the peptide (nonapeptide) seemed to be of importance, since either adding an arginine to, or removing an arginine from the C-terminus considerably affected the binding. However, replacing the C-terminal arginine with alanine or with D-arginine had little effect. Thus, it appears that the peptide bond between residues 8 and 9 contributed to binding, while the side chain of the amino acid at position 9 was probably not involved. In view of these data the authors raised the possibility that the peptide may exist in a cyclic form, and that alterations in the length of the peptide may adversely affect the stability of the ring.

In contrast to the length of the molecule, the overall molecular charge did not contribute to binding, nor did the local charge either at the N-terminal or the C-terminal arginine. Also, since phenylalanine in position 8 or serine in position 6 could be replaced by alanine without significantly affecting binding, it was concluded that changes from hydrophilic to hydrophobic groups, or vice versa, had little effect.

Changes in the bradykinin molecule which caused a dramatic decrease of binding with anti-bradykinin were those which affected the backbone structure of the peptide: a) substitution of glycine in position 4 by alanine, which leads to loss of the native conformation due to the limitation of rotation imposed by the β carbon; b) substitution of proline in position 2 by D-proline or by alanine; c) substitution of proline in position 3 or 7 by alanine or by D-proline; d) simultaneous replacement of the 3 prolines (positions 2, 3, 7) by D-proline. These data signified the importance of all of the proline residues (especially of proline in position 3), and led to the conclusion that a change in the overall shape of the peptide profoundly affected binding with antibody. This, coupled with the observation that peptide length was important, and with a molecular model of bradykinin which indicated that the molecule is of globular structure, strongly suggested that the antigenic determinant comprised the entire peptide, and that antibodies recognized the peptide in a preferred globular conformation (HABER et al., 1967).

The relationship between structure and the biological activity of bradykinin (on the estrus rat uterus) did not always correlate with the relationship between structure and antigenic activity (SPRAGG et al., 1967). However, both activities were profoundly affected by changes in the backbone of the peptide.

Gastrin

10
GLU–GLY–PRO–TRP–LEU–GLU–GLU–GLU–GLU–GLU–ALA–TYR–
17
GLY–TRP–MET–ASP–PHE (Human gastrin, BENTLEY et al., 1966).

Gastrin is a peptide hormone consisting of 17 amino acids. The natural gastrin molecule is termed gastrin I; synthetic gastrin (ANDERSON et al., 1964) is termed gastrin II, the latter differing from gastrin I by sulfation of the tyrosyl residue in position 12. Gastrin molecules from all of the species investigated, including man, have very similar amino acid sequences and differ from each other by only one or two amino acids (DAYHOFF and ECK, 1969). Moreover, the C-terminal pentapeptide amide of gastrin is present in the hormone of all the animal species investigated, and is responsible for the physiological activity of the peptide (GREGORY, 1966).

Antibodies to gastrin have been evoked in rabbits by immunization with synthetic human gastrin I (sequences 2–17) conjugated to protein carriers. These antibodies were demonstrated to bind with the radioiodinated homologous antigen, with gastrin molecules obtained from several other mammalian species (porcine, canine, ovine), and were shown to inhibit the physiological activity of the various gastrins (MCGUIGAN, 1968a). These observations are not surprising in view of the finding that the antibodies were directed against the C-terminal tetrapeptide amide. Moreover, caerulein and cholecystokinin-pancreozymin (both physiologically active peptides) which have the same C-terminal pentapeptide amide as gastrin (GREGORY, 1966; MUTT and JORPES, 1967; ANASTASI et al., 1968); also reacted with antibodies to gastrin (MCGUIGAN, 1969a, b).

Since the C-terminal tetrapeptide amide constitutes the antigenic determinant of gastrin, the tetrapeptide amide conjugated to a protein carrier was used for immunization of rabbits. The resulting antibodies reacted with synthetic human gastrin, with naturally occurring porcine gastrin, and with the C-terminal penta- and tetrapeptides of gastrin. Moreover, the antibodies were able to neutralize the hormonal activity of the molecules (MCGUIGAN, 1967). The antibodies produced to the C-terminal tetrapeptide amide conjugate reacted equally well with the C-terminal tetrapeptide as with the entire gastrin molecule, indicating that the tetrapeptide resided in an exposed position on the surface of the intact hormone, and furthermore, that the conformation of the gastrin molecule did not affect the spatial properties of the tetrapeptide. Interestingly, antibodies produced by immunization with synthetic human gastrin (2–17) conjugate bound the hormone cholecystokinin-pancreozymin, but substantially less well than they did with gastrin. On the other hand, antibodies elicited in response to immunization with the conjugate of the C-terminal tetrapeptide amide reacted equally well with intact gastrin and with cholecystokinin-pancreozymin (MCGUIGAN, 1968b). This, together with the finding that the carboxyl terminal tetrapeptide of gastrin reacted considerably less well than gastrin with antibodies to synthetic human gastrin (2–17) conjugate implies that determinants other than just the C-terminal tetrapeptide amide are present on the molecule. However, the fact that peptide 1–13 did not exhibit demonstrable binding argues against such determinants. Peptide 1–13 however, apparently does participate in the antigen-antibody interaction. Whether or not this participation is specific and depends upon the intact sequence of peptide 1–13 is not clear: On the one hand this peptide may enhance the binding of the C-terminal tetrapeptide amide in a nonspecific manner. On the other hand, the reduced reactivity of cholecystokinin pancreozymin with antibodies to gastrin conjugate (compared to the binding with the homologous peptide) may indicate some degree of specificity for peptide 1–13.

Ferredoxin

10
ALA–TYR–LYS–ILE–ALA–ASP–SER–CYS–VAL–SER–CYS–GLY–ALA–
CYS–ALA–SER–GLU–CYS–PRO–VAL–ASN–ALA–ILE–SER–GLN–GLY–
30
ASP–SER–ILE–PHE–VAL–ILE–ASP–ALA–ASP–THR–CYS–ILE–ASP–
50
CYS–GLY–ASN–CYS–ALA–ASN–VAL–CYS–PRO–VAL–GLY–ALA–PRO–
VAL–GLN–GLU (*Clostridium pasteurianum* ferredoxin according to TANAKA et al., 1966).

Interesting immunochemical investigations were performed on bacterial ferredoxins, which are relatively small, iron-containing proteins of molecular weight of approximately 5800, composed of 55 amino acids. Ferredoxin of *Clostridium pasteurianum* or its performic acid-oxidized form incorporated in FREUND's complete adjuvant, were immunogenic in rabbits. When native ferredoxin was used as an immunogen, the antibodies reacted much better with ferredoxin from which the iron was removed (oxidized ferredoxin) than with the homologous antigen. This phenomenon implies that prior to the initiation of the immune response the iron may have been removed *in vivo*. The reaction of these antibodies with various derivatives of ferredoxin showed that the 8 cysteine residues (these do not form disulfide bridges in the native molecule) played only a minor role in antigenicity (NITZ et al., 1969).

Since the peptides comprising residues Lys 3–Phe 30 and Val 31–Glu 55 were able to inhibit the reaction between performic acid–oxidized ferredoxin and homologous antibodies, and since the regions between Cys 8–Cys 18 and Cys 37–Cys 47 were apparently not antigenic (MITCHELL et al., 1970), it was deduced that peptides Lys 3–Ser 7, Pro 19–Phe 30, Val 31–Thr 36 or Pro 48–Glu 55 could contain antigenic determinants. Consequently, peptide Ala 1–Ser 7 (having the sequence Ala–Tyr–Lys–Ile–Ala–Asp–Ser), peptide Val 31–Thr 36 (having the sequence Val–Ile–Asp–Ala–Asp–Thr), and peptide Pro 48–Glu 55 (having the sequence Pro–Val–Gly–Ala–Pro–Val–Gln–Glu) were synthesized and assayed for antigenic activity. Of these three, the N-terminal heptapeptide and the C-terminal octapeptide exhibited significant activity (MITCHELL et al., 1970; KELLY and GERWING LEVY, 1971). Moreover, ferredoxin of *C. butyricum*, which differs from that of *C. pasteurianum* at position 9, exhibited strong reactivity with antiserum to the latter and to its performic acid-oxidized derivative. It is therefore interesting that a synthetic peptide representing the C-terminal octapeptide of *C. butyricum* ferredoxin (differing from the analogous peptide of *C. pasteurianum* ferredoxin in having Asn instead of Val in position 53) reacted poorly with antibodies to *C. pasteurianum* ferredoxin but reacted well with antibodies to performic acid-oxidized ferredoxin, in fact, as well as the homologous peptide. Further investigations (Christensen et al., 1972), showed that significant antigenic activity of the C-terminal octapeptide of ferredoxin resided in its C-terminal tetrapeptide portion having the sequence Pro–Val–Gln–Glu, and that marginal activity was exhibited even by the C-terminal tripeptide. The C-terminal pentapeptide (Ala–Pro–Val–Gln–Glu) and the entire C-terminal octapeptide exhibited equal binding with antibodies to performic acid-oxidized ferredoxin.

More recently these workers conjugated the N-terminal heptapeptide, the C-terminal tetrapeptide, or both peptides to a protein carrier. Each conjugate was shown to be reactive with antibodies to oxidized ferredoxin. Moreover, antibodies induced by immunization with the conjugates exhibited reactivity not only with the homologous peptides but also with oxidized ferredoxin (Teather and Gerwing Levy, 1971).

Cytochrome C

10
GLY–ASP–VAL–GLU–LYS–GLY–LYS–LYS–ILE–PHE–ILE–MET–
LYS–CYS–SER–GLN–CYS–HIS–THR–VAL–GLU–LYS–GLY–GLY–LYS–
30
HIS–LYS–THR–GLY–PRO–ASN–LEU–HIS–GLY–LEU–PHE–GLY–ARG–
50
LYS–THR–GLY–GLN–ALA–PRO–GLY–TYR–SER–TYR–THR–ALA–
ALA–ASN–LYS–ASN–LYS–GLY–ILE–ILE–TRP–GLY–GLU–ASP–THR–
70
LEU–MET–GLU–TYR–LEU–GLU–ASN–PRO–LYS–LYS–TYR–ILE–PRO–
GLY–THR–LYS–MET–ILE–PHE–VAL–GLY–ILE–LYS–LYS–LYS–GLU–
90
GLU–ARG–ALA–ASP–LEU–ILE–ALA–TYR–LEU–LYS–LYS–ALA–THR–
ASN–GLU (Human cytochrome C, according to Matsubara and Smith, 1962).

Cytochrome C is a protein of molecular weight of 12400, the primary structure of which is known for many species (Dayhoff and Eck, 1969). Possibly due to its small size and the retention of much of the primary structure throughout the phylogenetic scale, cytochrome C is usually only weakly immunogenic in rabbits. To enhance immunogenicity, it is commonly used in polymeric form. However, human cytochrome C was found to be immunogenic in its monomeric form (Reichlin et al., 1968a). A high degree of immunogenicity can also be achieved when cytochrome C is conjugated to carrier proteins. When tested with the monomeric homologous antigen, precipitating antibodies were found in antisera to human, monkey, and horse cytochromes C, but not in antisera to tuna or turkey cytochrome C (STAVITSKY, 1954; MARGOLIASH et al., 1970). However, antibodies to the latter did form precipitins and fixed complement when reacted with the polymerized homologous antigen. Excess of the monomeric form could completely inhibit the precipitin reaction and complement fixation reaction of polymers with the antisera, indicating the absence of polymer-specific determinants (MARGOLIASH et al., 1967a, b).

Although antibodies to any of 6 cytochromes C used as immunogens (human monkey, horse, tuna, kangaroo, and turkey) reacted to varying degrees with the cytochromes of 25 species, species-specific determinants have been demonstrated. The non-species-specific antibodies were in general directed against antigenic determinants present in all of the cytochromes of the cross-reacting species since there was no additivity in activity with antibodies when 2 cross-reacting proteins were reacted simultaneously. Cytochromes C having identical primary structures invariably turned out to be identical immunologically, but the reverse was not always true. Some of the sera did not distinguish between cytochromes C with different amino acid sequences, probably because the differences were not present in antigenic regions, or were not in regions which affect the conformation of the molecule (MARGOLIASH et al., 1967a, b; 1970). Antibodies from rabbits immunized with several cytochromes C were found to react with rabbit cytochrome C. Also, rabbits immunized with rabbit cytochrome C conjugated to bovine gamma globulin produced antibodies capable of binding autologous cytochrome C (NISONOFF et al., 1969).

Circular dichroism and optical rotatory dispersion analyses could not distinguish differences in structure of several cytochromes C (MARGOLIASH and SCHEJTER, 1966). Therefore, it is reasonable to assume that the serological distinction of various cytochromes C is related to differences in amino acid sequence. Studies with cytochromes C which differ in only one amino acid residue resulted in the localization of certain regions within the sequence as contributing to antigenicity. One such study involved human, horse, and donkey cytochrome C, which differ at position 47; horse cytochrome has a threonine residue, while human and donkey cytochromes have serine in this position. Rabbit anti-horse cytochrome C could not distinguish between horse and donkey cytochrome C, whereas rabbit anti-human cytochrome C was capable of distinguishing between these two cytochromes. It was concluded that the side chains of serine and of threonine affected the geometry of the determinant but did not actually participate in binding. Thus antibodies which were able to conform to a threonine-containing site (1 methyl group larger than serine) could accommodate the smaller serine residue but not vice versa (MARGOLIASH et al., 1967a, b). An alternative explanation may be offered, namely, that this area was immunosilent when horse cytochrome C was used as an immunogen in the rabbit, whereas when the human protein was used the antigenicity of the area was expressed. In another study, human and monkey (*Macaca mulatta*) cytochromes C, differing at position 58 (human having an isoleucine residue, monkey having a threonine residue) were compared by using rabbit antibodies (Nisonoff et al., 1970). Antiserum to the human protein was shown to contain 40% more antibodies reactive with the homologous than with the heterologous antigen. Antiserum to horse cytochrome C (with threonine at position 58) was more reactive with monkey than with human cytochrome C. Absorption of anti-human cytochrome C with an excess of monkey cytochrome C yielded a serum containing antibodies to a single determinant which was termed the "isoleucine site" (antibodies directed to this isoleucine site account for 30–40% of the total antibodies produced to human cytochrome C in rabbits). Only those cytochromes C which have an isoleucyl residue at this position (human and kangaroo) reacted with these specific antibodies. In contrast, antibodies to monkey cytochrome C reacted identically with monkey and human cytochromes C. Therefore, the presence of the hydrophobic isoleucine correlates with the occurence of an immunogenic determinant, whereas in the absence of isoleucine, this region is immunogenically silent (Nisonoff et al., 1970). Notwithstanding the above considerations it is also possible that the isoleucine is actually not part of an antigenic determinant, but rather that it is responsible for the induction of a determinant in another area of the chain.

Studies using Fab fragments from anti-human and anti-monkey cytochrome C supported the hypothesis of a single antigenic difference between the two molecules; with Fab from anti-human serum, each mole of monkey cytochrome C was found to bind 4 moles of Fab. As expected, no difference was observed with Fab from antibodies to monkey cytochrome C: both monkey and human cytochromes C bound 3 moles of Fab per mole protein (Williams and Reichlin, 1968). Work on the kinetics and the stoichiometry of the reaction between cytochrome C and antibodies also led to the conclusion that the number of antigenic sites on cytochrome C is small (Noble et al., 1969).

An interesting immunological finding was that ferricytochromes C (oxidized forms) from horse, okapi, cow, and kangaroo fixed more complement (10–15%) with some anti-horse cytochrome C sera than did the corresponding ferrocytochromes C (reduced forms) (Margoliash et al., 1967b). These findings confirm conclusions obtained by chemical studies that there is a conformational difference between the ferro- and ferri-forms of the molecule. When cytochrome C is used as an immunogen, it is most probably in the ferri-form.

Lysozyme

10
LYS–VAL–PHE–GLY–ARG–CYS–GLU–LEU–ALA–ALA–ALA–MET–

LYS–ARG–HIS–GLY–LEU–ASP–ASN–TYR–ARG–GLY–TYR–SER–

30
LEU–GLY–ASN–TRP–VAL–CYS–ALA–ALA–LYS–PHE–GLU–SER–

ASN–PHE–ASN–THR–GLN–ALA–THR–ASN–ARG–ASN–THR–ASP–

50
GLY–SER–THR–ASP–TYR–GLY–ILE–LEU–GLN–ILE–ASN–SER–

70
ARG–TRP–TRP–CYS–ASN–ASP–GLY–ARG–THR–PRO–GLY–SER–
ARG–ASN–LEU–CYS–ASN–ILE–PRO–CYS–SER–ALA–LEU–LEU–SER–
90
SER–ASP–ILE–THR–ALA–SER–VAL–ASN–CYS–ALA–LYS–LYS–ILE–
110
VAL–SER–ASP–GLY–ASP–GLY–MET–ASN–ALA–TRP–VAL–ALA–TRP–
ARG–ASN–ARG–CYS–LYS–GLY–THR–ASP–VAL–GLN–ALA–TRP–ILE–
ARG–GLY–CYS–ARG–LEU (Hen egg white lysozyme, Canfield, 1963).

Lysozyme is a single chain polypeptide of 129 amino acids. In addition to the amino acid sequence, the three-dimensional configuration of the enzyme molecule has been established through X-ray crystallography (Blake et al., 1965).

Several antigenic determinants of hen egg white lysozyme have been characterized by several workers (Fujio et al., 1959, 1962, 1968a, b; Shinka et al., 1962, 1967a, b). Two large peptides which exhibited haptenic activity encompass the amino acid sequence 57–107 (area I), and sequences 1–27 and 122–129 (area II). Disulfide bonds between cysteines 6 and 127, 64 and 80, and 76 and 94 were intact in these fragments. Of the total precipitable antibodies contained in a pool of anti-lysozyme serum, 47% of the antibodies were directed to area II and 12% to area I. Since the presence of one peptide did not influence the binding of the other peptide with anti-lysozyme, it was assumed that the antiserum contained at least 2 different populations of antibody molecules (Fujio et al., 1968b). Other investigations showed that peptic digestion of lysozyme yielded an immunologically active peptide consisting of residues 64–83 and 91–107 linked by two disulfide bridges at positions 64 and 80, and 76 and 94 (Canfield and Liu, 1968). Recently it has been shown that within the large peptide (residues 57–107) the area confined by residues 60–83, with a disulfide bridge between cysteine 64 and cysteine 80 (the "loop" peptide), was an antigenic determinant of native lysozyme (Arnon, 1968; Arnon and Sela, 1969; Maron et al., 1971). Antibodies to the "loop" peptide have been obtained by purification of anti-lysozyme with "loop" peptide immunoadsorbent. Moreover, immunization with "loop" peptide conjugated to poly D, L-alanyl-poly L-lysine produced antibodies which reacted with the "loop" peptide as well as with native lysozyme. The reduced and alkylated "loop" peptide did not bind with anti-lysozyme serum, but did bind appreciably with anti-"loop" peptide sera. These latter findings indicate that in antilysozyme serum antibodies were directed to a conformation-dependent determinant or determinants, whereas the anti-"loop" serum may have contained antibodies recognizing sequential determinants (Arnon and Sela, 1969).

Several groups of workers, using different animal species, have shown that the completely reduced and alkylated lysozyme (CM-lysozyme) and the native form of the protein exhibited minimal cross-reactivity, if at all (Shinka et al., 1967a; Gerwing and Thompson, 1968; Young and Leung, 1970). Furthermore, several peptides obtained by enzymatic digestion of CM-lysozyme did not inhibit the reaction between native hen egg white lysozyme and its homologous antiserum but did appreciably inhibit the reaction between CM-lysozyme and homologous antibodies. Thus, a determinate area of the native molecule may involve areas encompassing more than just linear sequences, i.e. areas dependent on the intact three-dimensional structure of the molecule. The CM-lysozyme determinate areas which were characterized consisted of sequences 1–12 and 106–129 (part of lysozyme area II) (YOUNG and LEUNG, 1970), and of sequence 74–96 (part of area I) (GERWING and THOMPSON, 1968). It thus appears that the determinate areas of lysozyme and of CM-lysozyme overlap. Further work is required for ascertaining whether lysozyme and CM-lysozyme share common determinants. An interesting finding was that the N-terminal peptide (residues 1–12) could appreciably inhibit the binding of the C-terminal peptide with anti-CM-lysozyme, indicating that the N- and the C-terminal regions in the CM-lysozyme molecule may have been in steric proximity to one another (YOUNG and LEUNG, 1970).

Studies with chemically modified derivatives of hen egg white lysozyme revealed certain amino acid residues and certain areas of the molecule which participated in reactivity with antibodies to the native molecule. A cyanogen bromide derivative of hen egg white lysozyme which had the disulfide bonds intact, but cleaved on the carboxyl side of the methionine residues at positions 12 and 105, was still capable of inhibiting the reaction between lysozyme and anti-lysozyme by as much as 70%, indicating that much of the conformation of the native molecule was retained (BONAVIDA et al., 1969). N-bromosuccinylation of tryptophan at positive 62 or 63 resulted in a derivative which could still completely inhibit the reaction between lysozyme and anti-lysozyme. However, extensive modification of all 6 tryptophan residues with 2-nitrophenyl-sulfonyl chloride completely destroyed the ability of the derivative to react with anti-native lysozyme. This modified lysozyme was also found to be very weakly immunogenic when compared to native lysozyme (HABEEB and ATASSI, 1969). In contrast, when all six lysine residues were succinylated, the derivative still inhibited as much as 42% of the reaction between lysozyme and anti-lysozyme (BONAVIDA, 1968). Similarly, modification of all six lysines with maleic or citraconic anhydride resulted in a molecule still possessing approximately 50% activity as compared with the native molecule (HABEEB and ATASSI, 1970).

The C-terminal peptide obtained from cyanogen bromide (CNBr) cleavage of CM-lysozyme was not able to inhibit the reaction of native lysozyme with anti-native lysozyme. However, this peptide could inhibit the reaction between native lysozyme and antibodies raised in rabbits in response to immunization with CNBr treated lysozyme. These observations led to the conclusion that there were determinants on hen lysozyme, called "heterotopes" to which antibodies were not formed when the native molecule was used for sensitization, but which were expressed when immunization was performed with a derivative, or with lysozyme from another species, which differs in this particular area (BONAVIDA, 1969; BONAVIDA and SERCARZ, 1971).

Extensive immunochemical comparisons have been performed on lysozymes from different origins. These comparisons led to the immunological prediction of sequence differences among the proteins (ARNHEIM and WILSON, 1967; ARNHEIM et al., 1969; MARON et al., 1970). In general, when the native molecules were compared, there was a good correlation between differences in the amino acid sequences and immunological distinction. However, although no immunological cross-reactivity between native human lysozyme and native hen egg white lysozyme could be demonstrated, the reduced and alkylated derivatives of these lysozymes significantly cross-reacted (ARNHEIM et al., 1971).

Myoglobin

10
VAL–LEU–SER–GLU–GLY–GLU–TRP–GLN–LEU–VAL–LEU–HIS–
20
VAL–TRP–ALA–LYS–VAL–GLU–ALA–ASP–VAL–ALA–GLY–HIS–GLY–
30
GLN–ASP–ILE–LEU–ILE–ARG–LEU–PHE–LYS–SER–HIS–PRO–GLU–
40 50
THR–LEU–GLU–LYS–PHE–ASP–ARG–PHE–LYS–HIS–LEU–LYS–
60
THR–GLU–ALA–GLU–MET–LYS–ALA–SER–GLU–ASP–LEU–LYS–
70
LYS–HIS–GLY–VAL–THR–VAL–LEU–THR–ALA–LEU–GLY–ALA–ILE–
80
LEU–LYS–LYS–LYS–GLY–HIS–HIS–GLU–ALA–GLU–LEU–LYS–PRO–
90
LEU–ALA–GLN–SER–HIS–ALA–THR–LYS–HIS–LYS–ILE–PRO–ILE–
110
LYS–TYR–LEU–GLU–PHE–ILE–SER–GLU–ALA–ILE–ILE–HIS–VAL–
LEU–HIS–SER–ARG–HIS–PRO–GLY–ASN–PHE–GLY–ALA–ASP–ALA–
130
GLN–GLY–ALA–MET–ASN–LYS–ALA–LEU–GLU–LEU–PHE–ARG–
150
LYS–ASP–ILE–ALA–ALA–LYS–TYR–LYS–GLU–LEU–GLY–TYR–GLN–
GLY (Sperm whale myoglobin, EDMUNDSON, 1965).

Myoglobin is an oxygen-carrying protein of molecular weight of 18400. In addition to its amino acid sequence, the three-dimensional structure of the molecule has been elucidated (KENDREW et al., 1961). Having no disulfide bridges, a great deal of its structural integrity resides in the high content of secondary structure which the molecule possesses; some 75 % of its residues are found in a helical configuration (KENDREW et al., 1961). A great deal of its antigenic activity can be attributed to its primary and secondary structures. Nevertheless, the tertiary structure of the molecule has also been implicated in antigenicity. It has been shown that antibodies to the heme-containing protein could detect differences between the heme-protein and the apo-form (REICHLIN et al., 1963). It has also been reported that while antibodies to metmyoglobin (the heme-containing protein with ferric iron) could not detect differences between metmyoglobin and apomyoglobin (without heme), the reactions of antibodies to apomyoglobin with metmyoglobin released the heme from the latter. In addition, it has been demonstrated that anti-apomyoglobin sera contained at least two populations of antibodies. One population could not distinguish between apomyoglobin and metmyoglobin, while the other could. Addition of heme to an immune precipitate, consisting of the latter group of antibodies and apomyoglobin, resulted in only partial reformation of metmyoglobin, whereas with the former group of antibodies complete reformation was observed (CRUMPTON, 1966).

A series of artificial myoglobins, containing various metals in the porphyrin, were reacted with antibodies to metmyoglobin. It was found that whereas the antigenic activity of artificial myoglobin with a copper metalloporphyrin was identical to that of the heme protein, all other derivatives showed diminished reactivity, which was attributed to conformational reorganizations (ATASSI, 1967a). Additionally, it was shown that a derivative of apomyoglobin, in which the tryptophan residues at positions 7 and 14 were modified (causing profound conformational effects), had drastically reduced antigenic activity compared to native apomyoglobin (ATASSI and CARUSO, 1968).

The importance of primary and secondary structures in the antigenicity of myoglobin was indicated by the fact that chymotryptic peptides could partially inhibit the reaction between anti-myoglobin and anti-apomyoglobin, and their respective antigens (CRUMPTON, 1964). Assay for immunological activity of several purified chymotryptic peptides revealed that five active peptides were adjacent to, or included in "corners" of the native molecule. A sixth active peptide constituted the C-terminal heptapeptide portion of the molecule. The N-terminal peptide, which is helical in the intact molecule, was inactive (CRUMPTON and WILKINSON, 1965). It was subsequently reported that the C-terminal heptapeptide (having the sequence Lys–Glu–Leu–Gly–Tyr–Gln–Gly) or even the C-terminal hexapeptide were active, but not as much as a peptide comprising the C-terminal 22 amino acids, the latter being predominantly helical in the native molecule. Subsequent experiments, using shorter peptides, revealed that the N-terminal lysine of the C-terminal heptapeptide was not involved in binding, but that one or both of the C-terminal amino acids, as well as the 4th and 5th amino acids from the C-terminus, were involved (CRUMPTON, 1967a, b). Studies with several synthetic C-terminal peptides indicated that the hepta- and hexapeptides were of equal reactivity with anti-myoglobin, whereas the penta- and tetrapeptides were progressively weaker. Analogues which contained phenylalanine or p-methoxy-phenylalanine in place of the tyrosine were as active as their native counterparts. It was concluded that the leucine residue, situated at a "corner" in the intact molecule, was essential for antigenicity, and furthermore, that the determinant was no smaller, and may be no larger, than a hexapeptide (CRUMPTON et al., 1970).

Several tryptic peptides of myoglobin, occupying several corners in the three-dimensional configuration of the molecule, as well as the C-terminal hexapeptide, were reactive with anti-myoglobin. Removal of the N-terminal heptapeptide from the molecule had no effect on its antigenic activity (ATASSI and SAPLIN, 1968). Cyanogen bromide cleavage of myoglobin at its 2 methionine residues yielded the antigenically active fragments 1–55, 56–131, and 132–153, the first two of which precipitated with anti-myoglobin and therefore contained 2 or more determinants. Removal of the N-terminal heptapeptide from fragment 1–55 reduced, but did not abolish, its precipitating activity. Thus, this N-terminal heptapeptide probably serves to stabilize the structure of the rest of the fragment. Of the three fragments, the one comprising residues 56–131 appeared to be the most active (ATASSI and SAPLIN, 1968). Modification of arginines 31 and 45 in fragment 1–55, as well as modification of arginine 139 in fragment 132–153, had no effect on antigenic activity. In contrast, modification of Arg 118 in the fragment 56–131 had an appreciable effect. This latter reduction in reactivity correlated well with the reduction in reactivity in the unfragmented molecule when it was modified at arginines 31, 118 and 139, suggesting that of these three arginines, Arg 118 was the only one involved in the antigenic activity of the intact molecule (ATASSI and THOMAS, 1969).

Cleavage of myoglobin at the proline residues yielded peptides encompassing residues 1–36, 37–87, 1–87, 37–119, and 120–153. Peptide 1–36 has been shown to precipitate with anti-apomyoglobin and had an activity equal to that of the cyanogen bromide fragment 1–55, suggesting that residues 37–55 were not involved in antigenicity (ATASSI and SINGHAL, 1970). Peptide 1–31 has been shown to be as reactive with antibodies to myoglobin as peptide 1–36. Succinylation of Lys 16 of peptide 1–31 abolished the reactivity, whereas carboxymethylation of the histidine residues at positions 12 and 24 had no effect (PERLSTEIN and ATASSI, 1971). Peptide 37–87 had been shown to inhibit the reaction between myoglobin and anti-myoglobin, and with a given antiserum the sum of the activities of peptides 1–36 and 37–87 was equal to that of fragment 1–87. Fragment 37–119 was also shown to have precipitating activity, but peptide 120–153 had only inhibitory activity. This activity of the latter was equal to that of the cyanogen bromide fragment 132–153 suggesting that residues 120–131 were inactive (ATASSI and SINGHAL, 1970).

As has already been indicated, studies with chemically modified derivatives have yielded valuable data concerning the antigenicity of myoglobin. It has been reported that totally acetylated myoglobin did not react with anti-myoglobin, neither by complement fixation nor by precipitin reaction. It was pointed out that the loss of activity could have been due to a conformational change, and not necessarily due to the loss of amino groups. In addition, treatment of myoglobin with carboxypeptidase removed the C-terminal glutamine and partially removed a pen ultimate serine, neither of which appeared to be important to antigenic activity (ATASSI, 1966). It is probable that this was a maverick myoglobin since the reported C-terminal dipeptide of sperm whale myoglobin is Gln–Gly (rather than Ser–Gln). It is interesting that the C-terminal dipeptide (Ser–Gln) of the protein investigated by ATASSI did not appear to play a role in antigenicity, whereas CRUMPTON demonstrated that the C-terminal dipeptide Gln–Gly (of normal sperm whale myoglobin) was essential for antigenicity (CRUMPTON, 1967a).

Other studies with chemically modified myoglobins have indicated that its two methionine residues, as well as tryptophan 7, were not involved in the antigenic activity of the protein (ATASSI, 1967b, 1969; ATASSI and CARUSO, 1968). Nitration of all three tyrosine residues of myoglobin diminished its antigenic activity (ATASSI, 1968). Nitration of the cyanogen bromide fragment 132–153 (which nitrated tryosines 146 and 151) abolished its reactivity.

Ribonuclease

10
LYS–GLU–THR–ALA–ALA–ALA–LYS–PHE–GLU–ARG–GLN–HIS–
20
MET–ASP–SER–SER–THR–SER–ALA–ALA–SER–SER–SER–ASN–
30
TYR–CYS–ASN–GLN–MET–MET–LYS–SER–ARG–ASN–LEU–THR–
40
LYS–ASP–ARG–CYS–LYS–PRO–VAL–ASN–THR–PHE–VAL–HIS–
50 60
GLU–SER–LEU–ALA–ASP–VAL–GLN–ALA–VAL–CYS–SER–GLN–
70
LYS–ASN–VAL–ALA–CYS–LYS–ASN–GLY–GLN–THR–ASN–CYS–
80
TYR–GLN–SER–TYR–SER–THR–MET–SER–ILE–THR–ASP–CYS–
90
ARG–GLU–THR–GLY–SER–SER–LYS–TYR–PRO–ASN–CYS–ALA–
100
TYR–LYS–THR–THR–GLN–ALA–ASN–LYS–HIS–ILE–ILE–VAL–ALA–
110 120
CYS–GLU–GLY–ASN–PRO–TYR–VAL–PRO–VAL–HIS–PHE–ASP–

ALA–SER–VAL (Bovine RNase, SMYTH et al., 1963).

Bovine pancreatic ribonuclease (RNase-A) has been extensively characterized (GROSS and WITKOP, 1962; KARTHA et al., 1967; SCHERAGA, 1967). Another form of the enzyme, RNase-B, has been isolated from natural sources and appears to differ from RNase-A in the replacement of an amide group with a carboxyl group (TANFORD and HAUENSTEIN, 1965). Treatment of RNase-A with subtilisin results in cleavage of the peptide bond between residues 20 and 21, yielding RNase-S, which is still enzymatically active. The activity is lost when the N-terminal eicosapeptide (S-peptide) is removed, resulting in a molecule termed S-protein (RICHARDS and VITHAYATHIL, 1959).

Rabbit antibodies to RNase-A could not distinguish between the A and the B form of the enzyme (BROWN et al., 1959a). Also, "early" antibodies to bovine RNase-A did not react with porcine RNase-A, although they were highly reactive with ovine RNase-A. However, antisera from later bleedings contained antibodies which reacted with bovine, ovine, and porcine RNase-A (BROWN et al., 1960). Boiling, or treatment of RNase-A with urea, had no effect on its binding with antibodies. Whereas alkaline denaturation reduced its antigenic activity to some extent, performic acid oxidation completely abolished immunological activity (BROWN et al., 1959a, b). However, some binding between oxidized RNase-A and antibodies to the native form was later demonstrated (BROWN et al., 1967). The reverse reaction (binding between native RNase-A and anti-oxidized RNase-A), could not be detected (BROWN, 1963). It may be significant that those antibodies to native RNase-A which were reactive with oxidized RNase-A were induced with the aid of FREUND's complete adjuvant or with alum precipitated antigen. Thus the presence of denatured forms in the native immunogen cannot be ruled out. Partially reduced RNase-A retained some reactivity with antibodies to the native form (BROWN et al., 1959b). This may be due to the retention of some of the native conformation, since it has been shown (NEUMANN et al., 1967) that RNase in which two disulfide bridges were open still retained its original conformation. A later study of the reaction between reduced and randomly re-oxidized RNase-A with antibodies to the native form (again induced with complete FREUND's adjuvant) demonstrated some reactivity, which was attributed to determinants of a size which would not be affected by conformational alterations (MILLS and HABER, 1963).

Studies on the reaction between antibodies to native RNase-A with RNase-S and with RNase-S protein demonstrated that, compared to RNase-A, RNase-S had diminished activity, the activity of the S-protein was even more reduced, and the RNase-S-peptide was inactive (SINGER and RICHARDS, 1959; MERIGAN and POTTS, 1966). Although RNase-A reacted with anti-RNase-S, it was shown that RNase-A lacked some determinants present in RNase-S, and that denaturation of RNase-S and the S-protein resulted in loss of reactivity with these antibodies. These findings may be significant when considered in conjunction with the three-dimensional models of RNase-A and of RNase-S (KARTHA et al., 1967; and WYCKOFF et al., 1967). The major difference between these two proteins is that in RNase-S the C-terminus of the S-peptide (residue 20) and the N-terminus of the S-protein (residue 21) have moved out and away from each other, and assumed a random coil configuration. It is thus probable that at least one antigenic determinant of RNase-A is located in the area of the peptide bond between residues 20 and 21. It is improbable that the determinant involves much of the S-peptide in light of its inability to inhibit the reaction between RNase-A and homologous antibodies.

Studies with other derivatives of RNase-A have alluded to the possible location of two other determinants on the molecule. It has been reported that deamination of the α-amino group of lysine 1 resulted in a loss of some antigenic activity (BROWN et al., 1959b; VAN VUNAKIS et al., 1960). It was further reported that RNase-A with the α amino group dinitrophenylated exhibited diminished reactivity, whereas guanidination of the ε amino of the same N-terminal lysine had no effect on reactivity with antibodies. In addition, removal of the N-terminal lysine had a large effect on reactivity, and removal of the N-terminal dipeptide (lysyl-glutamic acid) had an even greater effect (Brown et al., 1959b). Since the suggestion was made that the glutamic acid at position 2 was involved in the binding of the S-peptide to the S-protein (Finn and Hofmann, 1965) it is possible that this glutamic acid residue was either directly involved in binding with antibodies or was involved in stabilizing the N-terminal portion of the chain, or both. Another possible determinant has been suggested by studies with carboxypeptidase degraded RNase (Brown et al., 1959b). It was found that removal of the C-terminal valine had no effect, but that removal of the C-terminal tetrapeptide (Asp–Ala–Ser–Val) did diminish the reactivity with antibodies.

The reactivity of antibodies to oxidized RNase with several other denatured forms of the molecule indicated that the cysteine, but not the methionine residues, were important for antigenicity (May and Brown, 1968). The reaction between oxidized RNase-A and homologous antibodies has been shown to be inhibited by peptides 38–61 and 105–124 (containing 13 out of a total of 17 non-polar residues). Removal of amino acids 38 and 39 from peptide 38–61 greatly diminished, but did not eliminate reactivity (Brown, 1962). Additionally, peptide 105–124 has been found to strongly inhibit the reaction between antibodies to oxidized RNase-A and RNase-A which had been cleaved at the methionine residues with cyanogen bromide (Brown, 1963). More recently this peptide (105–124) has been found to bind directly with antibodies to oxidized RNase-A (Isagholian and Brown, 1970). Acetylation of peptide 38–61 at residues 41 and 61 has been shown to reduce its inhibitory activity by 80% (Brown, 1963). Since dinitrophenylation of the ε amino group of lysine 41 of oxidized-RNase did not affect its reactivity with antibodies (Brown et al., 1967), it is probable that the modification of residue 61 contributed to the lowered reactivity of the acetylated peptide.

Studies with antibodies which neutralize the enzymatic activity of RNase-A have yielded some interesting results. It has been shown that high affinity antibodies were effective inhibitors of enzymatic activity, and that neutralization was more effective in the presence of large substrates than of small substrates. This suggests that neutralization was effected as a result of steric hindrance by antibodies (Branster and Cinader, 1961). However, it was found that not all antibodies to RNase-A were neutralizing antibodies. Moreover, it was demonstrated that neutralizing and non-neutralizing antibodies could compete with each other to some extent, possibly due to their reactivity with overlapping sites on the RNase-A molecule (Cinader and Lafferty, 1963), or due to conformational alterations which can be induced by one or the other type of antibodies, leading to the exclusion of determinants recognized by the other antibodies. Indeed, such alterations have been shown to take place. An example of a conformational alteration by antibodies is that in which antibodies produced by one rabbit were shown to induce an increase in enzymatic activity (Cinader, 1967). This activation was found to be more pronounced with low molecular weight substrates. With high molecular weight substrases activation reached a peak and then fell off, suggesting that neutralizing antibodies were still able to bind to the activated enzyme. Additionally, it was found that the presence of substrate in the active site of the enzyme prevented the binding with the activating antibodies. Since the binding with the substrate is known to induce a conformational change in the enzyme (Sela et al., 1957), and in view of the fact that neutralizing antibodies could bind with the activated enzyme, it was concluded that the activating antibodies were binding to a site or sites distant from the active enzyme site, and were holding the molecule in a conformation which was more favorable for action on substrate. More recently, antibodies capable of increasing the activity of RNase-A on low molecular weight substrates were found in a chromatographic fraction of neutralizing antibodies (Suzuki et al., 1969). It was subsequently shown that this fraction could compete with neutralizing antibodies (Pelichova et al., 1970). It is probable that these antibodies bind at a site adjacent to the active site of the enzyme.

Tobacco Mosaic Virus and Its Protein Coat

10
SER–TYR–SER–ILE–THR–THR–PRO–SER–GLN–PHE–VAL–PHE–
LEU–SER–SER–ALA–TRP–ALA–ASP–PRO–ILE–GLU–LEU–ILE–ASN–
30
LEU–CYS–THR–ASN–ALA–LEU–GLY–ASN–GLN–PHE–GLN–THR–
50
GLN–GLN–ALA–ARG–THR–VAL–VAL–GLN–ARG–GLN–PHE–SER–
GLN–VAL–TRP–LYS–PRO–SER–PRO–GLN–VAL–THR–VAL–ARG–
70
PHE–PRO–ASP–SER–ASP–PHE–LYS–VAL–TYR–ARG–TYR–ASN–
ALA–VAL–LEU–ASP–PRO–LEU–VAL–THR–ALA–LEU–LEU–GLY–
90
ALA–PHE–ASP–THR–ARG–ASN–ARG–ILE–ILE–GLU–VAL–GLU–
ASN–GLN–ALA–ASN–PRO–THR–THR–ALA–GLU–THR–LEU–ASP–
110
ALA–THR–ARG–ARG–VAL–ASP–ASP–ALA–THR–VAL–ALA–ILE–
130
ARG–SER–ALA–ILE–ASN–ASN–LEU–ILE–VAL–GLU–LEU–ILE–ARG–
GLY–THR–GLY–SER–TYR–ASN–ARG–SER–SER–PHE–GLU–SER–
150
SER–SER–GLY–LEU–VAL–TRP–THR–SER–GLY–PRO–ALA–THR

(Strain *Vulgare*, Anderer et al., 1960; Tsugita et al., 1960).

The tobacco mosaic virus (TMV) is a rod like particle of molecular weight 40×10^6 having the approximate dimensions of 180×3000 Å. The virus is composed of 95% protein and 5% RNA. The protein component consists of

2200 identical protein subunits (TMVP), each subunit having a molecular weight of approximately 17500 and consisting of 158 amino acids. The structure of TMV and some immunological aspects of the virus and of the viral protein have been reviewed (KNIGHT, 1961; RAPPAPORT, 1965).

Two antigenic systems of TMV have been studied in detail: a) the intact virus, and b) the protein subunit (dissociated from the RNA core).

a) The Intact Virus as an Antigen

The C-terminal threonine residue of each protein subunit can be removed by treating the virus with carboxypeptidase. The infectivity of the virus thus treated (CT virus) remained unchanged while its antigenic properties have apparently been altered: anti-TMV absorbed with CT virus still precipitated with virus (but not with CT virus) whereas anti-CT virus absorbed with the intact virus still precipitated CT virus (but not with TMV) (HARRIS and KNIGHT, 1955).

Assays of the capacity of various peptides of the protein (obtained by the action of various enzymes) to inhibit the reaction between the virus and anti-virus revealed several active peptides, but all of which were rather inefficient inhibitors. Of these, the most efficient inhibitors (although causing only partial inhibition even when used simultaneously) were peptides comprising sequences 18–23, 62–68, 123–134, 129–131, 142–158, and 153–158, the latter representing the C-terminal portion of the protein and having the amino acid sequence Thr–Ser–Gly–Pro–Ala–Thr (ANDERER, 1963a). The fact that several peptides were capable of partially inhibiting the precipitin reaction between the virus and anti-virus suggested that antibodies to the virus were directed against several areas on the virus and implied that these areas were present on the exposed surface of the protein assembled on the viral RNA. Antisera to a synthetic antigen, composed of the C-terminal hexapeptide conjugated to a protein carrier, precipitated the virus (although the conjugate failed to precipitate with anti-virus (ANDERER, 1963b). Moreover, antisera prepared against the C-terminal penta-, tetra-, and tripeptides, each conjugated to a carrier protein, were able to inactivate the viral infectivity, the inactivating efficiency of the serum increasing with increases in the length of the peptide in the immunogenic conjugate (ANDERER and SCHLUMBERGER, 1965).

Although several peptides were obtained which were capable of inhibiting the reaction between TMV and anti-TMV (albeit to a low degree and at very high peptide to antigen ratio), there is considerable evidence indicating that the tertiary structure of the protein is of great importance in the antigenicity of TMV (ANDERER and HANDSCHUH, 1962, 1963; VON SENGBUSCH, 1965; VAN REGENMORTEL, 1966). Moreover, the orderly assembly of the protein subunits on the core RNA, resulting in the viral super structure, leads to the appearance of new determinants. These conformation-dependent determinants may be created by the steric proximity of residues from neighbouring subunits, or by the "packing" of protein subunits around the RNA which may result in a conformational change in individual protein subunits. Thus, new determinant(s) may consist of an area(s) or a conformation of an individual protein subunit in the aggregate, or of an area(s) composed of neighbouring sequences belonging to two or more subunits. In this connection, it has recently been shown that polymerization of TMV protein subunits (without RNA) brings about drastic changes in the antigenic determinants of the protein (RAPPAPORT and ZAITLIN, 1970).

b) The Protein subunit as an Antigen

Whereas the aforementioned peptides, and in particular the C-terminal portion of the protein, were found to constitute antigenic determinants of the virus (in the TMV anti-TMV system) they are of minor importance (if any) as determinant(s) of the isolated viral protein. A tryptic digest of TMVP was able to inhibit, to a large extent, the fixation of complement by TMVP and anti-TMVP. Further characterization revealed that the activity of the digest resided in the tryptic peptide comprising residues 93–112 of the protein, having the sequence Ile–Ile–Glu–Val–Glu–Asn–Gln–Ala–Asn–Pro–Thr–Thr–Ala–Glu–Thr–Leu–Asp–Ala–Thr–Arg (BENJAMINI et al., 1964). The activity of the digest was ascertained by several other methods, including the direct binding of (^{14}C)-N-acetyl peptide with antibodies to TMVP (BENJAMINI et al., 1965).

Experiments on the localization of the antigenic determinant on the eicosapeptide, utilizing the stepwise removal of N-terminal amino acids, showed that the binding with anti-TMVP did not appreciably diminish when 5 amino acid residues were removed (YOUNG et al., 1966). Moreover, the synthetic C-terminal decapeptide (having the sequence Thr–Thr–Ala–Glu–Thr–Leu–Asp–Ala–Thr–Arg) was demonstrated to constitute the antigenic portion of the eicosapeptide (STEWART et al., 1966). Further experiments, consisting of assaying the antigenic activity of C-terminal peptides varying in length from dipeptide to the decapeptide, indicated that the shortest peptide with specific binding with anti-TMVP was the C-terminal pentapeptide (having the sequence Leu–Asp–Ala–Thr–Arg) (YOUNG et al., 1967). This peptide could inhibit the reaction between (^{14}C)-N-acetyl decapeptide and anti-TMVP, indicating that the pentapeptide was the sole determinant area of the decapeptide. Moreover, since the decapeptide was able to inhibit the reaction between (^{14}C)-N-acetyl eicosapeptide and anti-TMVP, it was concluded that the C-terminal pentapeptide was the sole determinant of the eicosapeptide (BENJAMINI et al., 1968a). A synthetic decapeptide, consisting of the C-terminal pentapeptide to which five alanine residues were attached N-terminally, exhibited the same degree of binding with anti-TMVP as did the native decapeptide, suggesting that the enhanced binding with increased peptide length was apparently due to nonspecific factors.

Assays of closely related synthetic analogues of the C-terminal pentapeptide showed that the C-terminal tripeptide Ala–Thr–Arg was critical for antigenic specificity whereas the N-terminal dipeptide Leu–Asp was instrumental in expressing, or enhancing, the binding of Ala–Thr–Arg with anti-TMVP. By modifying the N-terminal residues, it was shown that a rough correlation existed between the hydrophobicity of the N-terminal portion of the pentapeptide and its binding with antibodies (YOUNG et al., 1968). Whereas binding of the tripeptide Ala–Thr–Arg with antibodies could never be demonstrated, (^{14}C)-N-octanoyl–Ala–Thr–Arg exhibited strong binding. In fact, more of the octanoylpeptide was found to bind with antibodies to TMVP than the native pentapeptide or the native decapeptide (BENJAMINI et al., 1968b). These data suggested that the antigenic specificity of the pentapeptide resided in its C-terminal tripeptide Ala–Thr–Arg, but that the binding with anti-TMVP was expressed through hydrophobicity N-terminal to this tripeptide.

While anti-TMVP sera produced by all the tested animals (which included rabbits, guinea pigs, mice, and sharks) reacted with the decapeptide having the sequence Thr–Thr–Ala–Glu–Thr–Leu–Asp–Ala–Thr–Arg, not all antisera reacted with the C-terminal pentapeptide. Antisera produced by some rabbits reacted with the C-terminal pentapeptide, sera produced by other rabbits reacted with the C-terminal hexapeptide, and sera produced by still other rabbits exhibited binding only with the C-terminal hepta- or octapeptide (BENJAMINI et al., 1968a). However, all the antisera tested contained antibodies which were shown to bind with N-octanoyl–Ala–Thr–Arg, indicating that immunization of all of the tested animals with TMVP led to the production, in all animals, of antibodies with specificity toward the tripeptide Ala–Thr–Arg.

The specificity, to Ala–Thr–Arg, of antibodies produced by different rabbits in response to immunization with the protein TMVP was further investigated by comparing the binding of antibodies with N-octanoyl–Ala–Thr–Arg and closely related octanoylated analogues (BENJAMINI et al., 1969). The analogues included the following octanoylated peptides: Gly–Thr–Arg, Leu–Thr–Arg, α-aminobutyryl–Thr–Arg, Ala–Gly–Arg, Ala–Ser–Arg, Ala–Leu–Arg, Ala–α-aminobutyryl–Arg, and α-Aminobutyryl-α-aminobutyryl–Arg. Of all these analogues, only N-octanoyl-α-aminobutyryl–Thr–Arg was found to bind with anti-TMVP produced by some, but not by all, rabbits. These findings point not only to the very high specificity of antibodies to Ala–Thr–Arg, but also show some subtle differences in the antibody specificity: of four rabbits tested, antibodies produced by one rabbit bound as much of the octanoyl α-aminobutyryl–Thr–Arg analogue as octanoyl Ala–Thr–Arg; those produced by another rabbit bound more of this octanoylated analogue than the octanoylated native sequence; those produced by another rabbit bound less, and those produced by still another rabbit could not be demonstrated to bind with this octanoylated analogue. These findings suggested that, in addition to possible differences in affinities, there also existed differences in specificities of the antibodies produced by the various rabbits in response to immunization with TMVP. Apparently, some antibody areas complementary to Ala–Thr–Arg fit the α-aminobutyryl analogue better than they fit the native sequence, while some areas fit the native sequence better than the analogue.

Attempts to immunize experimental animals with the antigenic peptides of TMVP were unsuccessful (SPITLER et al., 1970). Consequently, rabbits were immunized with the following peptides conjugated to succinylated bovine serum albumin: Thr–Thr–Ala–Glu–Thr–Leu–Asp–Ala–Thr–Arg, Leu–Asp–Ala–Thr–Arg, Leu–Asp–α-aminobutyryl–Thr–Arg, and Leu–Asp–Ala–Gly–Arg (the first three peptides are those which were previously shown to bind with anti-TMVP; antibodies with specificity to Leu–Asp–Ala–Gly–Arg could never be demonstrated in sera produced by over a dozen rabbits immunized with TMVP). Analysis of the specificity of antibodies induced by immunization with the different conjugates showed that antibodies induced by conjugates of those peptides which bind with anti-TMVP bound with such peptides, whereas antibodies produced by the conjugate of the peptide which does not bind with anti-TMVP bound only with the homologous peptide but with none of the peptides which were found to bind with anti-TMVP. Moreover, TMVP could specifically inhibit, to a significant extent, the reaction between anti-succinylated bovine serum albumin conjugated-Leu–Asp–Ala–Thr–Arg and the homologous peptide. In contrast, TMVP could not inhibit the binding of Leu–Asp–Ala–Gly–Arg and antibodies induced by Leu–Asp–Ala–Gly–Arg-conjugate (FEARNEY, 1970; FEARNEY et al., 1971).

The above results demonstrate that, in a general way, the relationship between structure of the peptide used for immunization and the specificity of the antibodies produced, parallels the relationship between peptide structure and binding with anti-TMVP (produced in response to immunization with the protein TMVP).

Insulin

A Chain

10

GLY–ILE–VAL–GLU–GLN–CYS–CYS–THR–SER–ILE–CYS–SER–

LEU–TYR–GLN–LEU–GLU–ASN–TYR–CYS–ASN.

B Chain

10

PHE–VAL–ASN–GLN–HIS–LEU–CYS–GLY–SER–HIS–LEU–VAL–

GLU–ALA–LEU–TYR–LEU–VAL–CYS–GLY–GLU–ARG–GLY–PHE–

30

PHE–TYR–THR–PRO–LYS–THR (Human insulin, according to NICOL and SMITH, 1960).

The insulin molecule consists of two polypeptide chains, an A chain having 21 amino acid residues, and a B chain having 30 residues. The chains are linked to each other by two disulfide bridges (Cys_A7–Cys_B7 and Cys_A20–Cys_B19). In addition, there is one intrachain disulfide bridge on the A chain between residues Cys_A6 and Cys_A11. Amino acid sequences of insulins from several sources are known (DAYHOFF and ECK, 1969).

The immunology of insulin has been extensively covered in several reviews (ARQUILLA, 1962; PROUT, 1962; POPE, 1966; SCHWICK, 1966; WILSON, 1967). Many mammalian insulins have been found to be immunologically cross-reactive. In addition, the antibodies neutralized the hormonal activities of most insulins. However, insulins of the coypu (DAVIDSON et al., 1968), the capybara (DAVIDSON et al., 1969), and the guinea pig (MOLONY and COVAL, 1955; KITAGAWA et al., 1960; DAVIDSON and HAIST, 1965) have been shown to be non-neutralizable by antibodies to beef insulin. Also the capybara insulin has been shown to be non-neutralizable by guinea pig antibodies to chicken and to cod insulins (DAVIDSON et al., 1969).

Human antibodies to beef insulin have been shown to be reactive with endogenous human insulin, although the interaction between beef insulin and these antibodies was stronger (GRODSKY, 1965). Antibodies of the IgE class, produced in humans in response to beef-pork therapeutic insulin, were shown to react with sheep and with human insulins but not with isolated A or B chains of these insulins (PATTERSON et al., 1969). In another report (BERSON and YALOW, 1959), it was shown that human anti-beef-pork therapeutic insulin was more reactive with beef and with sheep insulin than with pork or horse insulin, while human insulin was the least reactive. Since this order of reactivity parallels the known differences in amino acid sequence of these insulins, it was postulated that the region comprising residues 8–10 of the A chain (where the differences occur) may be involved in antigenicity. No differences were found between the binding of these antibodies with iodinated or with native insulin. Also, they could not distinguish between native and des-amido insulin, nor could they bind iodinated A or B chains. Since human anti-pork insulin (which differs from human insulin only in the C-terminal residue of the B chain) reacted with a derivative of pork insulin in which this amino acid was removed, it was concluded that the antibodies reacted with sequences of pork insulin which were common to human insulin (BERSON and YALOW, 1963).

Studies on the reactivity of human anti-beef insulin with various chemically modified derivatives of insulin showed that dinitrophenylated (lysines only), acetylated (85% of the available amino groups), desalanylated, sulfanylated, and methylesterified (at 6 out of 12 groups) insulins were almost identical to native insulin in their reactivity with antibodies. Methylesterified insulin (with 11 out of 12 groups), and oxidized insulin were only poorly reactive (GRODSKY et al., 1959).

Beef insulin which had been reacted with fluorescein isothiocyanate (another amino reactive compound) was less reactive with rabbit antibodies than native insulin (HALIKIS and ARQUILLA, 1961), although in a later report (ARQUILLA et al., 1966) it was indicated that the reactivity of this derivative with some antibodies was unchanged. A reduction in the antigenic activity of insulin which had been fluoresceinated at 0.9 groups per molecule has also been reported (BRUNFELDT et al., 1968). Iodination also reduced the antigenic activity of insulin, the loss of activity was proportional to the degree of iodination, although some activity remained even at 10 iodine atoms per molecule (BRUNFELDT et al., 1968). Other studies on the effect of iodination on the antigenic activity of insulin showed that iodination with even less than 1 atom per molecule resulted in reduced activity. These findings, coupled with earlier findings on iodinated derivatives of insulin indicate that the C-terminal portion of the A chain played an important role in the antigenicity and biological activity of the molecule (ARQUILLA et al., 1968).

Several derivatives of insulin (iodinated, fluoresceinated and desoctapeptide-insulin) were used to study the response of strain 2 and 13 guinea pigs to insulin (ARQUILLA et al., 1969). It was concluded that strain 2 guinea pigs produced antibodies which were predominantly directed against areas involving the N-termini of both the A and B chains, while strain 13 guinea pigs produced antibodies directed predominantly against a region involving the C-termini of both chains.

Studies with the isolated A and B chains of insulin have demonstrated that, whereas anti-A chain reacted only with the homologous antigen (isolated A chain), antibodies to the B chain reacted with both B chain and with intact insulin (YAGI et al., 1965; VARANDINI, 1967). Furthermore, anti-insulin was found to react with isolated B chain (YAGI et al., 1965). Subsequently it was reported that fragments representing residues B1–22, B1–16, and B1–10 exhibited reactivity with most antisera to B chain, while fragment B23–30 exhibited activity with some antisera only. The fragment B11–16 and B17–25 were inactive with all antisera tested (YAGI et al., 1967). Human antibodies to beef insulin were reactive with isolated B chain but not with A chain. Also, when two sheep antisera to sheep insulin and a pig antiserum to pig insulin were studied, it was found that the A chain had slight activity, the B chain had intermediate activity, and intact insulin was the most active (KERP et al., 1969). These latter studies, however, were carried out by measuring the displacement of iodinated insulin from antibodies, and the results of the experiments with the A chain should be interpreted with caution in light of the previously quoted studies (ARQUILLA et al., 1966, 1968) which indicated that A chain antigenicity was most readly affected by iodination.

The synthetic disulfide-linked dimers $B(1–8)_2$ and $B(17–30)_2$, were reactive with antibodies to insulin but $B(2–8)_2$, $A(13–21)_2$ and $A(10–21)_2$ were not (WILSON et al., 1967; KERP et al., 1968). In addition, fragments B1–8, B9–30, and B24–30 were reactive with anti-B chain, but $B(2–8)_2$, B9–14, and $B(17–23)_2$ were not. Also, studies on the reaction of A chain derivatives with anti-A demonstrated that A1–9 and A10–21 were reactive and that intact A chain could inhibit the insulin anti-insulin reaction (WILSON et al., 1967).

Studies on the cross-reaction between the A and B chains of beef, sheep, and pork insulins showed that the A chains of the beef and pork insulins (which differ in positions 8, 9 and 10) reacted similarly (but not quite identically) with anti-beed A chain, while sheep A chain was not so reactive. The three B chains (which are all identical in sequence) could not be distinguished by antibodies (VARANDINI, 1967). Studies on the antigenic activity of sulfonated, oxidized, or reduced A and B chains indicated that anti-S-sulfonated A chain was reactive with the homologous antigen only. Antibodies to oxidized or reduced A chains reacted with all three derivatives as well as with intact insulin, and with proinsulin. In addition, antibodies to any of the B chain derivatives reacted with all three derivatives, with insulin, and with proinsulin. Anti-insulin was shown to quantitatively bind oxidized B chain and 18% of oxidized A chain (TOUBER et al., 1970).

Investigations on the immunogenicity of insulin showed that 2 diabetic patients who were producing antibodies against beef insulin, which also reacted with human insulin, ceased to produce these antibodies when the administration of beef insulin was discontinued. This phenomenon suggests that human insulin was non-immunogenic although it was recognized by antibodies to beef insulin (GRODSKY, 1965). Other studies showed that when the B-30 residue (the C-terminal residue, which is the only residue in which pork, human, or rabbit insulins differ) was removed, the resulting molecule was immunogenic in each of the three species (LOCKWOOD and PROUT, 1962). These results indicate that the removal of the B30 residue from these insulins created new immunogenic determinants on these derivatives.

Hemoglobin

α Chain

10
VAL–LEU–SER–PRO–ALA–ASP–LYS–THR–ASN–VAL–LYS–ALA–

ALA–TRP–GLY–LYS–VAL–GLY–ALA–HIS–ALA–GLY–GLU–TYR–GLY–

30
ALA–GLU–ALA–LEU–GLU–ARG–MET–PHE–LEU–SER–PHE–PRO–

THR–THR–LYS–THR–TYR–PHE–PRO–HIS–PHE–ASP–LEU–SER–

50
HIS–GLY–SER–ALA–GLN–VAL–LYS–GLY–HIS–GLY–LYS–LYS–VAL–

70
ALA–ASP–ALA–LEU–THR–ASN–ALA–VAL–ALA–HIS–VAL–ASP–ASP–

MET–PRO–ASN–ALA–LEU–SER–ALA–LEU–SER–ASP–LEU–HIS–ALA–

90
HIS–LYS–LEU–ARG–VAL–ASP–PRO–VAL–ASN–PHE–LYS–LEU–LEU–

110
SER–HIS–CYS–LEU–LEU–VAL–THR–LEU–ALA–ALA–HIS–LEU–PRO–

ALA–GLU–PHE–THR–PRO–ALA–VAL–HIS–ALA–SER–LEU–ASP–LYS–

130
PHE–LEU–ALA–SER–VAL–SER–THR–VAL–LEU–THR–SER–LYS–

TYR–ARG

β Chain

10
VAL–HIS–LEU–THR–PRO–GLU–GLU–LYS–SER–ALA–VAL–THR–
ALA–LEU–TRP–GLY–LYS–VAL–ASN–VAL–ASP–GLU–VAL–GLY–
30
GLY–GLU–ALA–LEU–GLY–ARG–LEU–LEU–VAL–VAL–TYR–PRO–
TRP–THR–GLN–ARG–PHE–PHE–GLU–SER–PHE–GLY–ASP–LEU–
50
SER–THR–PRO–ASP–ALA–VAL–MET–GLY–ASN–PRO–LYS–VAL–
70
LYS–ALA–HIS–GLY–LYS–LYS–VAL–LEU–GLY–ALA–PHE–SER–ASP–
GLY–LEU–ALA–HIS–LEU–ASP–ASN–LEU–LYS–GLY–THR–PHE–
90
ALA–THR–LEU–SER–GLU–LEU–HIS–CYS–ASN–LYS–LEU–HIS–VAL–
110
ASP–PRO–GLU–ASN–PHE–ARG–LEU–LEU–GLY–ASN–VAL–LEU–
VAL–CYS–VAL–LEU–ALA–HIS–HIS–PHE–GLY–LYS–GLU–PHE–THR–
130
PRO–PRO–VAL–GLN–ALA–ALA–TYR–GLN–LYS–VAL–VAL–ALA–
GLY–VAL–ALA–ASN–ALA–LEU–ALA–HIS–LYS–TYR–HIS (Human hemoglobin, according to Braunitzer et al., 1961).

Hemoglobin is an oxygen-carrying molecule composed of two pairs of polypeptide chains each of which carries a heme group. It is of approximate molecular weight of 67000 (in mammals). In addition to the amino acid sequence, the three-dimensional structure of the molecule has been elucidated (Perutz, 1965). Studies on the immunochemical properties of porcine hemoglobin and horse muscle myoglobin demonstrated that rabbit antisera to hemoglobin or to myoglobin were able to distinguish between the heme proteins and their globin moieties, indicating conformation-dependent determinants in these molecules (heme does not directly participate in antigenicity) (Reichlin et al., 1963).

The reaction between antibodies to human hemoglobin and the oxyform of the molecule was stronger than with the deoxy-form, regardless of the form of the immunogen. The difference between the reactivity of the oxy- and the deoxy-form was most pronounced when these two forms of human hemoglobin reacted with antibodies to horse hemoglobin, suggesting that the discrimination between the conformational differences (of oxy- or deoxy-forms) was better when the antibodies were produced in response to immunization with a distantly related, but cross-reacting, hemoglobin (Reichlin et al., 1964a, 1965a). Hemoglobins containing various other ligands (such as CO) were indistinguishable from oxyhemoglobin. These observations support chemical evidence which indicates that deoxy-hemoglobin has a different conformation from liganded hemoglobins, especially with respect to the *β* chains (Muirhead and Perutz, 1963). Antibodies to native human oxyhemoglobin were able to distinguish between the oxy- and the deoxy-forms of two hybrids (containing 2 canine α chains and 2 human β chains or vice versa), but could not distinguish one oxy- or deoxy-hybrid from the other oxy- or deoxy-hybrid, respectively. These findings suggested that the α and β chains were equally involved in the conformational changes accompanying oxygenation (REICHLIN et al., 1965b).

Rabbit antibodies to human hemoglobin A_1 have been shown to distinguish mutant hemoglobins S and C (differing only in position 6 of the β chain), and hemoglobin H (composed of 4 β chains) from A_1 hemoglobin. The single amino acid difference at position 6 in hemoglobins A_1, S, and C may have altered the binding with antibody via an alteration in a local conformation in the N-terminus, although no profound differences were found between the crystal structures of A_1 and S oxyhemoglobin (PERUTZ et al., 1951). It is also possible that the change at position 6 was accompanied by a conformational change far removed from the N-terminus (REICHLIN et al., 1964b). When 18 other mutant human hemoglobins were tested with rabbit antisera to human hemoglobin A_1, the immunological activity of the mutant proteins was inversely related to the extent of differences between the primary sequence of human and rabbit hemoglobins. Six mutants, which differed from A_1 in their reactivity with anti-A_1, differed from A_1 in those regions where human and rabbit hemoglobins contain many differences in primary sequence. Mutants indistinguishable from A_1 by anti-A_1 differed in regions of the polypeptide where the primary sequence of human and rabbit hemoglobins are identical. Thus it was postulated that the region containing the differences in primary structure between rabbit and human hemoglobins were the ones to which rabbit antibodies were directed (REICHLIN, 1969).

Immunochemical experiments with anti-A_1 hemoglobin and its separated chains demonstrated that the intact molecule contained some determinants which could be attributed to the α and β chains. However, anti-A hemoglobin also contained antibody populations which were directed against quaternary structure-dependent determinants, formed by the interaction of α and β chains, regardless of the species from which the complementary chains were derived (REICHLIN et al., 1965a).

Rabbit antibodies induced by immunization with the isolated chains (α or β) reacted with the immunogen as well as with intact hemoglobin but did not react with the opposite chain. Antibodies to isolated α chains reacted with the homologous antigen in either the free form or when complexed with the β, indicating that the determinants of the α chains were exposed in hemoglobin. However these antibodies were less reactive with α chains when complexed with γ chains (fetal hemoglobin), indicating a structural difference(s) between adult hemoglobin ($\alpha_2 \beta_2$) and fetal hemoglobin ($\alpha_2 \gamma_2$) with respect to the α chain. The reaction between antibodies to the β chains and adult hemoglobin was poorer than with isolated β chains, indicating that some determinants of the β chains were lost when the β chains complexed with α chains to form $\alpha_2 \beta_2$ (REICHLIN et al., 1966).

Quantitative studies on the specificity of antibodies to human hemoglobins (REICHLIN, 1970) showed that fetal hemoglobin was much more immunogenic in rabbits than was adult hemoglobin. This was probably due to the increased immunogenicity of the γ chains, since the actual concentration of antibodies specific to α chains was equal in antisera to adult and to fetal hemoglobins. Also, in anti-adult hemoglobin there were equal portions of anti-α specific and anti-β specific antibodies, whereas in anti-fetal hemoglobin sera there was a 2–3 fold higher concentration of anti-γ specific antibodies. Since the number of radical differences in primary structure between human and rabbit α and β chains respectively are similar (7 differences between human α chains and rabbit β chains, and 11 differences between human and rabbit α chains) the antigenicity of both chains is also similar. However the human γ chain and rabbit β chain differ to a much greater extent (21 radical amino acid differences), therefore, the γ chain is more immunogenic in rabbits than the α or β chains. Nevertheless, the residues responsible for binding with antibodies may not be the same as those involved in immunogenicity, and furthermore, only certain regions may be involved in antigenicity.

Whereas isolated α or β chains of hemoglobin were reported to be immunogenic (REICHLIN et al., 1966; REICHLIN, 1970), the immunogenicity of the denatured α or β chains has also been reported (OVARY, 1964; KRAUS and SASSANO, 1965). However, although confirming the immunogenicity of denatured β chain, ASKONAS and SMYTH (1964) could not demonstrate the immunogenicity of the denatured α chain.

Studies on the reactivity of isolated tryptic peptides from the α chain with antibodies to denatured α chains indicated that the N-terminal area 1–7, area 93–99 and the C-terminal area 128–139 were antigenically active. In addition, when a longer tryptic peptide containing residues 1–11 was used, it was found to be more active than peptide 1–7. It was suggested that the longer peptide stabilized a preferred conformation, or possibly increased hydrophobic binding with antibody (KRAUS, 1967a, b).

Collagen

Unlike other proteins discussed in this review, the amino acid sequencing of collagen has thus far nor been completed. However, the immunochemical studies on collagen, especially those with synthetic "collagen-like" peptides, prompted the inclusion of this material in this chapter.

Available data indicate that except for 14 to 16 residues at the amino terminal end of the molecule, the amino acid sequence of collagen is highly repetitive, with the triplet Gly–Pro (or Hyp)-X being the repeating unit (for a summary see PIEZ et al., 1968). Sequence data of collagens from various species indicate that this basic structure is highly conserved. The universality of the Gly–Pro-X unit, or variation thereof, in the structure of collagen suggests that these residues may be responsible for the unique triple helical structure of the molecule. This led several investigators to synthesize peptides which possessed structural characteristics similar to collagen (for a review on the subject see CARVER and BLOUT, 1967).

Immunochemical information germane to the antigenicity of collagen and of gelatin (thermally denatured collagen) was obtained from studies with collagen, with gelatin, with random co-polymers, and with polymers and peptides related to collagen.

Various amounts of tyrosine were introduced onto the polypeptide chains of gelatin. Antibodies produced in response to immunization with gelatin enriched with tyrosine (2.4%) reacted well with gelatin whereas another immunogen, consisting of gelatin enriched with tyrosine (11%), induced the formation of antibodies of poor reactivity with gelatin. Gelatin derivatives enriched with L-tryptophan, L-lysine, and L-glutamic acid (to the extent of more than 20%), showed only slight cross-reactivity with gelatin (SELA and ARNON, 1960). The same workers synthesized the following polymers and co-polymers and tested their inhibitory effects on the reaction between gelatin and antigelatin: poly L-proline, poly L-hydroxyproline, random co-polymers of L-proline and glycine, L-proline and sarcosine, L-hydroxyproline and glycine, L-proline glutamic acid and glycine, and a low molecular weight polymer of the tripeptide Pro–Gly–Gly. None of the above substance had any inhibitorye ffect.

Studies on the antigenic properties of a group of synthetic polyimino acids demonstrated the antigenicity of the linear homopolymer, poly L-proline, in guinea pigs (JASIN and GLYNN, 1965a). This polymer exists in two forms differing from one another only in their secondary structure, form I has a right-handed helix and form II with a left-handed helix. Both forms were antigenic yet non-cross-reacting, an effect which emphasizes the importance of secondary structure in these systems. Antisera to the homopolymer poly-L-proline, or the random co-polymer of L-proline, L-hydroxyproline, and glycine, did not cross-react with collagen. These results are in agreement with those mentioned earlier (SELA and ARNON, 1960) which failed to demonstrate cross reactions between anti-collagen and randon co-polymers. The same workers (JASIN and GLYNN, 1965b) have studied the antigenic properties of the following synthetic peptides: poly hydroxy L-proline form II, poly acetyl-hydroxy L-proline form I, poly (L-proline, glycine, acetyl-hydroxy L-proline) form II, poly (L-proline, glycine hydroxy L-proline), poly (L-proline, glycine) form II, and poly L-proline form II. The immunological activity of the polymers was tested in guinea pigs and in rabbits by active cutaneous anaphylaxis, by passive agglutination, and by fluorescent antibody microscopy. All the tested peptides were found to be immunogenic in rabbits and guinea pigs except poly hydroxyproline and poly acetyl-hydroxyproline. The results on the antigenic relationships between the synthetic peptides, collagen, and acetylated collagen suggested that the acetyl-hydroxyproline formed the common antigenic determinant for both the acetylated co-polymers and acetylated collagen. However, no cross-reaction between the unacetylated peptide and calf skin collagen could be demonstrated by skin tests. Also, tanned red blood cells coated with a co-polymer of L-proline, L-hydroxyproline, and glycine failed to agglutinate when tested with anti-collagen sera of guinea pigs or rabbits. The authors concluded that, like in SELA and ARNON's study, the actual differences between the random co-polymers and collagen were great enough to make the two antigens immunologically non-cross-reacting.

The ordered polymer $(Pro–Gly–Pro)_n$ of a relatively high molecular weight was immunogenic in guinea pigs as shown by passive cutaneous anaphylaxis as well as by immediate and delayed skin sensitivity tests (SELA et al., 1967; BOREK, 1968). This polymer was also immunogenic in rabbits, as demonstrated by passive cutaneous anaphylaxis. The cross-reaction between anti-$(Pro–Gly–Pro)_n$ and fish collagen (Ichthyocol), rat skin collagen and guinea pig skin collagen, was ascertained by passive cutaneous anaphylaxis. All the reactions recorded as positive were of relatively weak intensity. The cross reaction between anti- $(Pro–Gly–Pro)_n$ and collagen indicated that the two materials shared antigenic determinants. Moreover, the cross-reaction of guinea pig skin collagen with guinea pig antibodies against $(Pro–Gly–Pro)_n$ demonstrated that it was possible to induce antibodies against autologous collagen with a synthetic antigen. Since $(Pro–Gly–Pro)_n$ was shown to exist in solution as a triple stranded helix of the collagen type (ENGLE et al., 1966), and since X-ray diffraction studies of the same polymer showed that it had a diffraction pattern similar to that obtained with collagen (TRAUB and YONATH, 1966), it was concluded that the cross-reaction between anti-$(Pro–Gly–Pro)_n$ and various collagens was a result of the close conformational similarity between this co-polymer and collagen (SELA et al., 1967).

In the search for determinants shared by collagen, gelatin and peptides, an octapeptide having the amino acid sequence Gly–Pro–Gly–Pro–Pro–Gly–Ala–Lys was synthesized and tested for immunological activity related to collagen. Rabbit antiserum against guinea pig skin collagen precipitated this peptide. Moreover, a tryptic digest of collagen could inhibit the reaction between the anti-collagen and the peptide. The specific binding of the peptide with anti-collagen was also confirmed by experiments on the direct binding of the (^{14}C) internally labelled peptide with anti-collagen sera obtained from two rabbits (KETTMAN et al., 1967). Moreover, antisera produced in rabbits in response to immunization with the peptide conjugated to succinylated bovine serum albumin were shown to react with collagens and/or gelatins obtained from skins of guinea pigs, mice, rats, rabbits, and humans (LEUNG et al., in preparation).

The studies with synthetic polypeptides have demonstrated that the best approach to mimmicking the antigenic activity of collagen is not through the use of random co-polymers of glycine, proline, and hydroxyproline, but rather through the use of ordered synthetic co-polymers or peptides of known sequences. Since both $(Pro–Gly–Pro)_n$ and the octapeptide synthesized by KETTMAN et al. exhibited cross-reaction with collagen, a comparison between the two may indicate the likely antigenic determinant(s) that are responsible for this cross-reaction. Thus, it is evident that the sequence Pro–Gly–Pro–Pro–Gly is shared by the ordered co-polymer $(Pro–Gly–Pro)_n$ and by the synthetic octapeptide. Recent studies on the activity of C-terminal areas of the octapeptide indicated that the activity of the latter was confined in its C-terminal pentapeptide having the sequence Pro–Pro–Gly–Ala–Lys (LEUNG et al., in preparation). Therefore, the sequence Pro–Pro–Gly is that which is shared with the co-polymer $(Pro–Gly–Pro)_n$, and may indeed constitute the determinant responsible for the immunological cross-reaction with collagen.

III. Discussion

Circulating antibodies elicited in response to immunization with protein antigens may be directed against one or more of the structural aspects of the protein. These consist of the primary structure (the sequential arrangement of amino acids in the polypeptide chain), the secondary structure (which is dictated by the backbone of the polypeptide, such as α helix and β structure), the tertiary structure (which is conferred by the interaction between various reactive groups on the polypeptide chain and is expressed by the folding of the chain), and, when applicable, the quaternary structure (brought about by the interaction of protein subunits to form a super structure). Discussions concerning the specificity of antibodies to these aspects of protein antigens have been the subject of several reviews (MAURER, 1964; SELA, 1966, 1969; SELA et al., 1967; KABAT, 1968; PRESSMAN and GROSSBERG, 1968; GOODMAN, 1969). Therefore only a few examples of the interaction of antibodies with various aspects of protein structure, drawn from the several proteins dealt with in the present work, will be discussed. The major portion of the present discussion will focus on the relationship between antigenic sequences and binding with circulating antibodies. The relationship between antigenic structure of some of the proteins dealt with in this chapter and immunological specificity at other levels of the immune response (i.e. delayed hypersensitivity, immunogenicity, etc.) are dealt with elsewhere (BENJAMINI et al., 1972).

Antibodies to Quaternary Structures

Immunochemical data on those protein antigens composed of more than one polypeptide chain or one protein subunit, clearly indicate that whereas the separated chains may be immunologically active, antisera elicited in response to immunization with the native protein contain some antibody populations which are directed against architectural aspects present only in the intact molecule. Thus, anti-TMV cannot be completely absorbed with the protein subunit (TMVP), and antibodies to highly aggregated forms of the protein subunit cannot be completely absorbed with lower aggregation forms. Moreover, detailed immunochemical work resulted in the characterization of two different immunodominant areas, one of which consists of a C-terminal hexapeptide portion of the protein and represents an antigenic determinant of the virus, and another which is confined in area 93–112 of the protein and represents a determinant of the protein subunit.

Immunochemical studies with hemoglobin showed that antisera to the native molecule A_1 hemoglobin ($\alpha_2 \beta_2$) contain antibody populations which react specifically with determinants formed as a result of the interaction between the α and β chains: whereas human α chain alone cannot fix complement with anti-A_1 human hemoglobin, complement fixation is achieved by a hybrid hemoglobin consisting of 2 human α chains and 2 canine β chains. The β chains apparently induce in the α chains a conformation which is similar to that which they possess in the native human hemoglobin, and to which some antibodies are directed. Also, it has been shown that some antibodies against the isolated α chains of hemoglobin recognize determinant(s) expressed when the α chains are either in the free state or complexed with β chains in the normal $\alpha_2 \beta_2$ tetramer, but not expressed when the α chains are complexed with γ chains to form $\alpha_2 \gamma_2$ (fetal hemoglobin).

Quaternary structure-dependent determinants may consist of areas formed by the juxtaposition of two or more interacting subunits. On the other hand they may consist of structures on a given subunit which are induced by the interaction with other subunits. Nevertheless, polymerization of a protein does not necessarily result in quaternary structure-dependent determinants, nor does it necessarily affect the antigenicity of the monomers. A case in point is cytochrome C. Studies with this protein revealed that a polymer of cytochrome C exhibits better binding with antibodies than does the monomer. However, since the monomer is able to completely inhibit the reaction between the polymer and antibodies induced in response to immunization with the polymer, it may be concluded that the serum contains no polymer-specific antibodies.

Antibodies to Tertiary Structures

A great proportion of antibodies induced by immunization with protein antigens is directed against the tertiary structure of the protein. This is evident from numerous experiments. For example, lysozyme cross-reacts poorly, if at all, with reduced and carboxymethylated lysozyme; performic acid-oxidized RNase cross-reacts only poorly, if at all, with native RNase. In general, the more extensive the alteration of the protein, the less cross-reactivity with antibodies to the native protein is witnessed. With a few exceptions, immunochemical studies with most of the proteins discussed in this chapter have demonstrated the presence of antibodies to conformation-dependent determinants. Antibodies are able to distinguish between the ferri- and ferro-forms of cytochrome C, confirming the demonstration by chemical methods that there exists a conformational difference between the two forms. Similarly, antibodies can distinguish between the oxy- and deoxy-forms of hemoglobin and myoglobin, forms which differ from one another in their conformation. Other examples are those of the antigenicity of myoglobin with metalloporphyrins other than heme, and the effects of modification of tryptophan 14 on the reactivity of myoglobin with antibodies to the native molecule. In all of these examples, conformational alterations parallel a reduction in reactivity with antibodies. Two other interesting examples relevant to tertiary structure-dependent determinants are those concerning the antigenicity of the N- and C-termini of reduced and carboxymethylated (CM) lysozyme, and some determinants of insulin. In both of these cases there appear to exist antibodies which are directed against areas formed by the juxtaposition of two distant regions of the main chain.

Antibodies to Secondary Structures

The exquisite specificity of antibodies is directed not only to the conformations of a polypeptide which are expressed by its folding, but also to those structural aspects which are governed by the polypeptide's backbone. The demonstration that antiserum to bradykinin contains antibodies which fail to bind with analogues of the peptide in which amino acid substitutions result in changes in the backbone structure of the molecule indicates the presence of antibodies which recognized those structural aspects of the peptide dictated by its secondary structure. Another example which indicates the presence of antibodies to secondary structure-dependent determinants is the immunological cross-reaction between several collagens and gelatins and the polymer $(Pro–Gly–Pro)_n$. All of these antigens form, in solution, triple-stranded helices. Results of immunochemical analyses of myoglobin, which revealed the existence of antigenic determinants in several "corners" of the molecule, are also indicative of secondary structure-dependent determinants.

The inhibitory capacity of fragments of a given protein on the reaction between the protein and homologous antibodies is usually limited since the fragment in solution most probably lacks the configuration which it has in the intact molecule, and since conformation-dependent antibodies may recognize determinants consisting of areas composed of different parts of the polypeptide chain. Nevertheless, the binding of small fragments with antibodies to the whole protein has been demonstrated for many globular proteins. On the one hand such fragments may represent sequence-dependent (rather than conformation-dependent) determinants. On the other hand, there also exists the possibility that a given sequence can transconform to fit the combining site of an antibody molecule which is actually directed against a conformation containing this fragment. Moreover, antibodies may be directed against areas consisting of different parts of the primary sequence, brought into proximity by the folding of the polypeptide chain. Consequently, it is conceivable that a sequence representing part of this area may bind with an antibody site which is actually directed against the whole area. An example for the latter is indicated from the immunochemical analysis of reduced and carboxymethylated lysozyme: this form of the molecule undoubtedly possesses a conformation different from that of the native molecule (there is little or no serological cross-reactivity between the two forms); however, the N-terminal and the C-terminal portions of the molecule, respectively, were demonstrated to bind with antibodies to the reduced and carboxymethylated form of the enzyme. The binding of antibodies with the N-terminal portion can be inhibited by the C-terminal portion, and vice versa, indicating that the immunogen has a preferred conformation with the N- and C-termini in close proximity, and that the antibodies are directed against an area consisting of both portions of the molecule.

Antibodies to Primary Structures

Sequence-dependent antigenic determinants have been demonstrated for many polypeptides and proteins. In fact, immunochemical studies with peptides of relatively low molecular weight, which are poor immunogens, if at all, but which become immunogenic when conjugated to carriers, showed that, in most cases, the antibodies induced react with amino acid sequences constituting various portions of the peptide. Thus, antibodies to ACTH, to gastrin, and to angiotensin, have been shown to react with determinants which depend upon their amino acid sequences rather than upon more complex architectural features of the peptide. Moreover, in these systems, the capacity of relatively short, isolated peptides to inhibit the reaction between the native peptide and antibodies is generally greater than that of isolated peptides from complex globular proteins. Nevertheless, relatively high antigenic activity was also found to be possessed by fragments of proteins of relatively high molecular weight (which are immunogenic by themselves), many of which are globular. Although the transconformation of a peptide to an antibody site actually directed against a conformation-dependent determinant cannot be ruled out, studies which show that immunization with the active peptide conjugated to protein carriers elicits antibodies, some of which are capable of reacting with the native protein, indicate that sequence-dependent determinants do indeed exist.

The major attribute of all antigenic determinants is their specificity. Subtle changes in the side groups of amino acids which make up the determinant affect, to a greater or lesser extent, its binding with antibodies; pronounced changes generally lead to the inactivation of antigenicity. Thus, the immunologically active tetrapeptide amide of gastrin becomes inactive when the peptide acquires a free C-terminal carboxyl group; change of the valine to asparagine in the antigenically active C-terminal octapeptide of *Clostridium pasteurianum* ferredoxin greatly reduces reactivity with antibodies; acetylation of lysine 61 in the immunologically active peptide comprising residues 38–61 of ribonuclease greatly reduces its reactivity with antibodies to oxidized ribonuclease. However, some changes in amino acid composition of antigenic peptides do not appreciably affect binding. This does not mean lack of antigenic specificity but rather may mean that although the amino acid in question constitutes part of a peptide which binds with antibodies, it is not essential to the complementary fit between the peptide and the antibody site. In fact, often the complete removal of such an amino acid residue will only minimally affect the binding.

The immunochemical work on an antigenic determinant of the tobacco mosaic virus protein may serve to illustrate some of the points just mentioned. The pentapeptide Leu–Asp–Ala–Thr–Arg binds with antibodies to the whole protein, whereas the C-terminal tetrapeptides is inactive. Work with analogues of the peptide demonstrated that substitution of the leucine by glycine, or of aspartic acid by glutamic acid results in inactive peptides, implying that these two N-terminal amino acids are involved in antigenic specificity of the peptide. Although in the true sense of the word "specificity" this interpretation seems correct, subsequent work has shown that the substitutions of these N-terminal amino acids with other amino acids possessing various degrees of hydrophobicity yield peptides with relative binding activities correlating with hydrophobicity of this N-terminal region. Moreover, substituting these two N-terminal amino acids with octanoic acid results in an immunologically active peptide with antigenic specificity confiend to the C-terminal tripeptide moiety. Further, it was shown that except for substitution of the alanine residue in the octanoylated peptide by α-aminobutyric acid, many other substitutions with closely related amino acids yield inactive peptides. The antigenically active native pentapeptide may thus be regarded as consisting of two functional areas: the C-terminal tripeptide which is critical for antigenic specificity, and the N-terminal area which enhances and expresses the binding of the C-terminal tripeptide.

It has been suggested that hydrophobic area(s) enhance binding through a hydrophobic interaction with antibodies (KARUSH, 1962; METZGER et al., 1963; SINGER, 1965; BENJAMINI et al., 1969). This implies that in addition to having an area complementary to the antigenic determinant Ala–Thr–Arg, the antibody combining site contains hydrophobic area(s) which interact with the hydrophobic area(s) of, or in proximity to the antigenic determinant. Since the hydrophobic area of the antigen does not require rigid structural restrictions (in the antigenic peptide of TMVP it may consist of leucine, isoleucine, tyrosine, pentaalanine, acetyl-, or octanoyl-), it is conceivable that its interaction with the hydrophobic area on the antibody is essentially nonspecific, and serves as an important auxillary force in the binding between the specific, sterically complementary area of the antigen (Ala–Thr–Arg) and the antibody site. However, a restriction may be imposed on this hydrophobic interaction, namely that the participating hydrophobic groups on the antigen and antibody, respectively, must be present in the correct position for interaction. This restriction is dictated by the juxtaposition of the sequence responsible for antigenic specificity and by the antibody area complementary to this sequence. The hydrophobic area of the antibody may be situated spacially at or near the site complementary to antigenic specificity and thus contribute to binding. This situation may be analogous to the suggested hydrophobic fold or slot located at or near the catalytic site of α-chymotrypsin or alcohol dehydrogenase, which contributes to the binding of normal substrates (ROYER and CANADY, 1968). It is thus possible to view the binding of the antigenic pentapeptide with antibodies as being governed by an area comprising its antigenic specificity and a hydrophobic area which enhances the binding.

Although, as has just been discussed, the antigenic specificity of the immunologically active peptide Leu–Asp–Ala–Thr–Arg is mainly governed by the C-terminal tripeptide portion, the N-terminal portion may play some role in specificity. The latter portion can be substituted by octanoic acid without loss of activity. However, substitution of the aspartic acid with glutamic acid yields an inactive pentapeptide Leu–Glu–Ala–Thr–Arg. It is improbable that the change results in a drastic decrease in hydrophobicity in this area. The reduced activity is probably due do the steric interference of the glutamic acid residue in the binding of Ala–Thr–Arg (with or without the N-terminal leucine) with antibodies.

The aforementioned model of a determinant and its binding with the complementary antibody site is not necessarily universal: other determinants may have both antigenic specificity and binding properties exhibited simultaneously by the same group of residues. Also, auxiliary binding forces other than hydrophobic may be of importance (such as hydrogen bonds, electrostatic interactions, etc.) (see KARUSH, 1962, and SINGER, 1965 for reviews). Nevertheless, the model is conceptually convenient, especially in view of accepted models for substrate-enzyme interactions.

It is difficult to conceive that an immunologically active peptide composed, for example, of 10 or 15 amino acid residues will contain more than one antigenic determinant. Indeed, most immunologically active peptides of this size or smaller do not react with antibodies as multivalent antigens (this however does not exclude the presence of more than 1 determinant and may result from the steric hinderance for the reactivity of 2 antibody molecules with a short peptide). Often, the chemical modification or removal of some residues results in either no loss, or in a decreased binding with antibodies, but not in the complete abolition of binding. Several examples for this phenomenon include the following: the binding of the whole gastrin molecule with antibodies to gastrin is substantially higher than that of the C-terminal tetrapeptide amide although the latter is the sole determinant of the molecule, and although gastrin from which this C-terminal peptide has been removed is incapable of binding. The antigenic activity of the C-terminal octapeptide of ferredoxin of *C. pasteurianum* is attributed to its C-terminal tetrapeptide; however, although the latter exhibits binding with antibodies the addition of alanine N-terminally to the peptide results in a marked increase in binding. A synthetic peptide of the sequence Lys–Glu–Leu–Gly–Tyr–Gln–Gly which is the C-terminal heptapeptide of sperm whale myoglobin was demonstrated to bind with antibodies to this protein; the activity of shorter C-terminal peptides with antibodies to myoglobin is in the order hepta- = hexa- > penta- >> tetrapeptide, clearly indicating that antigenic specificity resides in the

C-terminal tetra- or pentapeptide. It is possible that, in the latter example, the leucine contributes to the binding of the pentapeptide through increasing the hydrophobicity at this portion of the molecule. Returning to the determinant of the tobacco mosaic virus protein, the pentapeptide mentioned earlier is the C-terminal portion of an eicosapeptide representing sequence 93–112 of the protein; although the pentapeptide is the only determinant of the eicosapeptide, the binding with antibodies is in the order eicosapeptide > C-terminal decapeptide > C-terminal pentapeptide. In fact, the addition of 5 alanine residues N-terminal to the antigenic pentapeptide results in a peptide exhibiting binding similar in magnitude to that of the native decapeptide, although the N-terminal pentapeptide portion of the former differs drastically from the native sequence. These few examples serve to illustrate that, in addition to areas which are responsible for antigenic specificity, antigenically active peptides may contain sequences, which although not directly participating in antigenic specificity, may considerably affect binding with antibodies. This brings to light another point, namely, that even those determinants which depend upon structural features such as secondary, tertiary or quaternary structures may be considered to consist of areas which are primarily involved in antigenic specificity, and areas which enhance binding with antibodies.

By analogy to an antigenic determinant, the complemetary antibody site may be conceptually regarded as consisting of two functionally different (although not mutually exclusive) areas: one which is complementary to that portion of the determinant which governs specificity and another which contributes to enhance the binding with the determinant. If this is so, then the great variations in affinities of antibodies to a given antigenic determinant may be due not only to variations in specificities but also due to variations in the auxiliary forces involved in binding, thereby contributing to variations in affinities.

Size of Combining Sites

At this point in the discussion, the size of the antibody reactive site merits consideration. Since present technology does not allow the physicochemical characterization of an antibody reactive site, the only approach to estimations of the size is through careful evaluation of the size of antigenic determinants to which the antibody combining site is complementary.

An estimation of the size of the combining site of antibodies became possible following the classical work of KABAT, who, using the inhibitory activity of a series of oligosaccharides of increasing size on the precipitin reaction between dextran and human anti-dextran, estimated the dimensions of the combining site to be $34 \times 12 \times 7$ Å. Subsequently, several other estimations were made, all of which were similar to KABAT'S estimate. A combining site of such dimensions could accommodate optimally a group equivalent to the size of 5–8 amino acid residues (Reviewed by KABAT, 1966). Careful immunochemical work on the combining site size, using synthetic polypeptidyl antigens, revealed that the size was that which was able to accommodate 4 amino acid residues (SCHECHTER et al., 1966).

Analysis of the determinant size (and the complementary antibody site) of the protein antigens discussed herein is in agreement with the aforementioned dimensions. The size of the antibody combining site to poly γ-D-glutamic acid has been estimated to be $36 \times 12 \times 7$ Å (the size of a hexaglutamic acid). Antibodies to oxidized *C. pasteurianum* ferredoxin were demonstrated to bind with a determinant consisting of a tetrapeptide. Similarly, a tetrapeptide was demonstrated to bind with antibodies to gastrin. An antigenic determinant of TMV was found to consist of a hexapeptide. A determinant of TMV protein consist of a pentapeptide, and a pentapeptide or even a tetrapeptide was found to constitute one of the antigenic determinants of myoglobin.

Although tetra- and pentapeptides have been shown to bind with antibodies, care should be taken in interpreting the size of the antibody combining site. On the one hand a peptide of a given size may fill only partially the antibody site; on the other hand, binding may eventually be demonstrated even with shorter peptides. Consequently, the definition of the site is reduced to an operational term i.e. the ability to demonstrate binding by one method or another. This is true not only for the estimation of the lower size limit but also of the upper size limit. As was already mentioned, the binding of short peptides with antibodies can be increased by attaching various ligands (amino acid sequences or other ligands) either C-terminally or N-terminally to the peptide. In several instances, the attachment of as many as 10 or 15 additional residues to a peptide substantially increased binding with antibodies, although these additional sequences did not exhibit antigenic specificity.

Although the determination of the combining site size may be handicapped by operational factors, it is clear that antigenic specificity, and by analogy, the complementary antibody combining site, is governed by very few amino acid residues.

References

ANASTASI, A., ERSPAMER, V., ENDEAN, R.: Isolation and amino acid sequence of caerulein, the active decapeptide of the skin of *Hyla caerulea*. Arch. Biochem. **125**, 57–68 (1968).

ANDERER, F. A.: Versuche zur Bestimmung der serologischen determinanten Gruppen des Tabakmosaikvirus. Z. Naturforsch. **18**b, 1010–1014 (1963a).

ANDERER, F. A.: Preparation and properties of an artificial antigen immunologically related to tobacco mosaic virus. Biochim. biophys. Acta (Amst.) **71**, 246–248 (1963b).

ANDERER, F. A., HANDSCHUH, D.: Die Reihenfolge der Aminosäuren im Protein des Tabakmosaikvirus. IV. Spaltung des Tabakmosaikvirusproteins mit Trypsin. Z. Naturforsch. **17**b, 536–543 (1962).

ANDERER, F. A., HANDSCHUH, D.: Einfluß der Tertiärstruktur auf die serologischen Eigenschaften des Tabakmosaikvirus. Z. Naturforsch. **18**b, 1015–1021 (1963).

ANDERER, F. A., SCHLUMBERGER, H. D.: Properties of different artificial antigens immunologically related to tobacco mosaic virus. Biochim. biophys. Acta (Amst.) **97**, 503–509 (1965).

ANDERER, F. A., UHLIG, H., WEBER, E., SCHRAMM, G.: Primary structure of the protein of tobacco mosaic virus. Nature (Lond.) **186**, 922–925 (1960).

ANDERSON, J. C., BARTON, M. A., GREGORY, R. A., HARDY, P. H., KENNER, G. W., MACLEOD, J. K., PRESTON, J., SHEPPARD, R. C., MORLEY, J. S.: The antral hormone gastrin. II. Synthesis of gastrin. Nature (Lond.) **204**, 933–934 (1964).

ARNHEIM, N., PRAGER, E. M., WILSON, A. C.: Immunological prediction of sequence differences among proteins. J. biol. Chem. **244**, 2085–2094 (1969).

ARNHEIM, N., SOBEL, J., CANFIELD, R.: Immunochemical resemblance between human leukemia and hen egg white lysozyme and their reduced carboxymethyl-derivatives. In preparation (1971).

ARNHEIM, N., WILSON, A. C.: Quantitative immunological comparison of bird lysozymes. J. biol. Chem. **242**, 3951–3956 (1967).

ARNON, R.: A selective fractionation of anti-lysozyme antibodies to different determinant specificities. Europ. J. Biochem. **5**, 583–589 (1968).

ARNON, R., SELA, M.: Antibodies to a unique region in lysozyme provoked by a synthetic antigen conjugate. Proc. nat. Acad. Sci. (Wash.) **62**, 163–170 (1969).

ARQUILLA, E. R.: Immunological studies on insulin. Ciba Found. Coll. Endocrinol. **14**, 146–164 (1962).

ARQUILLA, E. R., BROMER, W. W., MERCOLA, D.: Immunology, conformation, and biological activity of insulin. Diabetes **18**, 193–205 (1969).

ARQUILLA, E. R., OOMS, H., FINN, J.: Genetic differences of combining sites of insulin antibodies and importance of C-terminal portion of the A chain to biological and immunological activity of insulin. Diabetologia **2**, 1–13 (1966).

ARQUILLA, E. R., OOMS, H., MERCOLA, D.: Immunological and biological properties of iodoinsulin labeled with one or less atoms of iodine per molecule. J. clin. Invest. **47**, 474–487 (1968).

ASKONAS, B. A., SMYTH, D. G.: Antigenicity of the β chain of human hemoglobin. Nature (Lond.) **201**, 496–497 (1964).

ATASSI, M. Z.: Role of the amino groups and C-terminal of sperm-whale myoglobin in the antigen-antibody reaction. Nature (Lond.) **209**, 1209–1211 (1966).

ATASSI, M. Z.: Immunochemistry of sperm-whale myoglobins prepared with various modified porphyrins and metalloporphyrins. Biochem. J. **103**, 29–35 (1967a).

ATASSI, M. Z.: Periodate oxidation of sperm-whale myoglobin and the role of the methionine residues in the antigen-antibody reaction. Biochem. J. **102**, 478–487 (1967b).

ATASSI, M. Z.: Immunochemistry of sperm-whale myoglobin. III. Modification of the three tyrosine residues and their role in the conformation and differentiation of their roles in the antigenic reactivity. Biochemistry **7**, 3078–3084 (1968).

ATASSI, M. Z.: Immunochemistry of sperm-whale myoglobin. V. Specific modification of the methionine residues with β-propiolactone. Immunochemistry **6**, 801–810 (1969).

ATASSI, M. Z., CARUSO, D. R.: Immunochemistry of sperm-whale myoglobin. II. Modification of the two tryptophan residues and their role in the conformation and antigen-antibody reaction. Biochemistry **7**, 699–706 (1968).

ATASSI, M. Z., SAPLIN, B. J.: Immunochemistry of sperm-whale myoglobin. I. The specific interaction of some tryptic peptides and of peptides containing all the reactive regions of the antigen. Biochemistry **7**, 688–698 (1968).

ATASSI, M. Z., SINGHAL, R. P.: Immunochemistry of sperm-whale myoglobin. VIII. Specific intraction of peptides obtained by cleavage at proline peptide bonds. Biochemistry **9**, 3854–3861 (1970).

ATASSI, M. Z., THOMAS, A. V.: Immunochemistry of sperm-whale myoglobin. IV. The role of the arginine residues in the conformation and differentiation of their roles in the antigenic reactivity. Biochemistry **8**, 3385–3394 (1969).

AXELROD, A. E., TRAKETELLIS, A. C., HOFMANN, K.: Antigenicity of synthetic polypeptides with known amino acid sequences. Nature (Lond.) **197**, 146–148 (1963).

BENJAMINI, E., SCIBIENSKI, R. J., THOMPSON, K.: The relationship between antigenic structure and immune specificity. In: Contemporary topics in immunochemistry, Vol. I, p. 149 (F. P. INMAN, and W. J. MANDY ed.). New York: Plenum Press 1972 (in press).

Benjamini, E., Shimizu, M., Young, J. D., Leung, C. Y.: Immunochemical studies on the tobacco mosaic virus protein. VI. Characterization of antibody populations following immunization with tobacco mosaic virus protein. Biochemistry **7**, 1253–1260 (1968a).

Benjamini, E., Shimizu, M., Young, J. D., Leung, C. Y.: Immunochemical studies on the tobacco mosaic virus protein. VII. The binding of octanoylated peptides of the tobacco mosaic virus protein with antibodies to the protein. Biochemistry **7**, 1261–1264 (1968b).

Benjamini, E., Shimizu, M., Young, J. D., Leung, C. Y.: Immunochemical studies on tobacco mosaic virus protein. IX. Investigations on binding and antigenic specificity of antibodies to an antigenic area of TMVP. Biochemistry **8**, 2242–2246 (1969).

Benjamini, E., Young, J. D., Peterson, W. J., Leung, C. Y., Shimizu, M.: Immunochemical studies on the tobacco mosaic virus protein. II. The specific binding of a tryptic peptide of the protein with antibodies to the whole protein. Biochemistry **4**, 2081–2085 (1965).

Benjamini, E., Young, J. D., Shimizu, M., Leung, C. Y.: Immunochemical studies on the tobacco mosaic virus protein. I. The immunological relationship of the tryptic peptides of tobacco mosaic virus protein to the whole protein. Biochemistry **3**, 1115–1120 (1964).

Bentley, P. H., Kenner, G. W., Sheppard, R. C.: Structures of human gastrins I and II. Nature (Lond.) **209**, 583–585 (1966).

Berson, S. A., Yalow, R. S.: Species-specificity of human anti-beef, pork insulin serum. J. clin. Invest. **38**, 2019–2025 (1959).

Berson, S. A., Yalow, R. S.: Antigens in insulin: determinants of specificity of porcine insulin in man. Science **139**, 844–845 (1963).

Blake, C. C. F., Koenig, D. F., Mair, G. A., North, A. C. T., Phillips, D. C., Sarma, V. R.: Structure of hen egg-white lysozyme. A three-dimensional Fourier synthesis at 2 Å resolution. Nature (Lond.) **206**, 757–761 (1965).

Bonavida, B.: The molecular basis of the serological specificity of hen egg white lysozyme. Ph. D. thesis, Univ. of Calif., Los Angeles (1968).

Bonavida, B.: Non-immunogenic epitopes on hen lysozyme. Fed. Proc. **28**, 326 (1969).

Bonavida, B., Miller, A., Sercarz, E. E.: Structural basis for immune recognition of lysozymes. I. Effect of cyanogen bromide on hen egg white lysozyme. Biochemistry **8**, 968–979 (1969).

Bonavida, B., Sercarz, E. E.: Structural basis for immune recognition of lysozymes. II. Reactive but non-immunogenic epitopes on hen lysozymes. Europ. J. Immunol. (1972) (in press).

Borek, F.: Delayed-type hypersensitivity to synthetic antigens. In: Current topics in microbiology and immunology, vol. 43, p. 126–161 (M. Sela, ed.). Berlin-Heidelberg-New York: Springer 1968.

Boyd, B. W., Peart, W. S.: The production of high-titer antibody against free angiotensin. II. Lancet **1968** II, 129–133.

Branster, M., Cinader, B.: The interaction between bovine ribonuclease and antibody. A study of the mechanism of enzyme inhibition by antibody. J. Immunol. **87**, 18–38 (1961).

Braunitzer, G., Gehring-Müller, R., Hilschmann, N., Hilse, K., Hobom, G., Rudloff, V., Wittmann-Liebold, B.: Die Konstitution des normalen adulten Humanhämoglobins. Hoppe-Seylers Z. physiol. Chem. **325**, 283–286 (1961).

Brown, R. K.: Antigenic structure of ribonuclease. III. Inhibition by peptides of antibody to performic acid-oxidized ribonuclease. J. biol. Chem. **237**, 1162–1167 (1962).

Brown, R. K.: Immunologic studies of bovine ribonuclease derivatives. Ann. N.Y. Acad. Sci. **103**, 754–764 (1963).

Brown, R. K., Delaney, R., Levine, L., van Vunakis, H.: Studies on the antigenic structure of ribonuclease. J. biol. Chem. **234**, 2043–2049 (1959a).

Brown, R. K., Durieux, J., Delaney, R., Leikhim, E., Clark, B. J.: Immunochemical reactions involving ribonuclease. Ann. N.Y. Acad. Sci. **81**, 524–541 (1959b).

Brown, R. K., McEwan, M., Mikoryak, C. A., Polkowski, J.: Studies on the antigenic structure of ribonuclease. V. Dinitrophenylated ribonuclease. J. biol. Chem. **242**, 3007–3013 (1967).

BROWN, R. K., TACEY, B. C., ANFINSEN, C. B.: The reaction of porcine and ovine ribonucleases with antibody to bovine ribonuclease. Biochim. biophys. Acta (Amst.) **39**, 528–530 (1960).

BRUNFELDT, K., HANSEN, B. A., JORGENSEN, K. R.: The immunological reactivity and biological activity of iodinated insulin. Acta endocr. (Kbh.) **57**, 307–329 (1968).

CANFIELD, R. E.: The amino acid sequence of egg white lysozyme. J. biol. Chem. **238**, 2698–2707 (1963).

CANFIELD, R. E., LIU, A. K.: The disulfide bonds of egg white lysozyme (muramidase). J. biol. Chem. **240**, 1997–2002 (1968).

CARVER, J. P., BLOUT, E. R.: Polypeptide models for collagen. In: Treatise on collagen, vol. **11**, p. 441–526 (G. M. RAMACHANDRAN, ed.). New York: Academic Press 1967.

CATT, K., COGHLAN, J. P.: Generation and use of antibodies to angiotensin. II. Aust. J. exp. Biol. med. Sci. **45**, 269–282 (1967).

CHRISTENSEN, J. A., GERWING LEVY, J., KELLY, B.: Immunochemical studies on peptides related to the C-terminal antigenic determinant of performic acid oxidized ferredoxin from *Clostridium pasteurianum*. Biochemistry (1972) (in press).

CINADER, B.: Antibodies to enzymes—a discussion of the mechanisms of inhibition and activation. In: Antibodies to biologically active molecules (B. CINADER, ed.), p. 85–137. Oxford: Pergamon Press 1967.

CINADER, B., LAFFERTY, K. J.: Antibody as inhibitor of ribonuclease: The role of steric hindrance, aggregate formation, and specificity. Ann. N.Y. Acad. Sci. **103**, 653–673 (1963).

CRUMPTON, M. J.: The immunological activity of chymotryptic peptides of sperm-whale myoglobin. Biochem. J. **91**, 4C–6C (1964).

CRUMPTON, M. J.: Conformation changes in sperm-whale metmyoglobin due to combination with antibodies to apomyoglobin. Biochem. J. **100**, 223–232 (1966).

CRUMPTON, M. J.: An antigenic site of sperm-whale myoglobin. Nature (Lond.) **245**, 17–20 (1967a).

CRUMPTON, M. J.: The molecular basis of the serological specificity of proteins, with particular reference to sperm-whale myoglobin. In: Antibodies to biologically active molecules (B. CINADER, ed.), p. 61–84. Oxford: Pergamon Press 1967b.

CRUMPTON, M. J., LAW, H. D., STRONG, R. C.: The C-terminal antigenic site of sperm-whale myoglobin: The immunological activities of synthetic peptides related to the C-terminus of myoglobin. Biochem. J. **116**, 923–925 (1970).

CRUMPTON, M. J., WILKINSON, J. M.: The immunological activity of some of the chymotryptic peptides of sperm-whale myoglobin. Biochem. J. **94**, 545–556 (1965).

D'AURIAC, G. A., MEYER, P., MILLIEZ, P.: Cross-reaction of angiotensin II with other angiotensin analogs. C. R. Acad. Sci. (Paris), Ser. D **268**, 1886–1888 (1969).

DAVIDSON, J. K., HAIST, R. E.: Failure of guinea pig antibody to beef insulin to neutralize pancreatic and serum guinea pig insulin activity *in vitro*. Canad. J. Physiol. Pharmacol. **43**, 373–378 (1965).

DAVIDSON, J. K., ZEIGLER, M., HAIST, R. E.: Failure of guinea pig antibody to beef insulin to neutralize coypu (nutria) insulin. Diabetes **17**, 8–12 (1968).

DAVIDSON, J. K., ZEIGLER, M., HAIST, R. E.: Failure of guinea pig antibodies to beef insulin, chicken insulin, and cod insulin to neutralize capybara insulin. Diabetes **18**, 212–215 (1969).

DAYHOFF, M. O., ECK, R. V.: Atlas of protein sequence and structure 1969. Silver Spring, Maryland: National Biomedical Research Foundation.

DEODHAR, S. D.: Immunologic production of antiangiotensin. II. Production and detection of antiangiotensin. J. exp. Med. **111**, 429–439 (1960).

DIETRICH, F. M.: Immunochemical analysis of rabbit antibodies against angiotensin II. Immunochemistry **4**, 65–76 (1967).

DIETRICH, F. M., FRISCHKNECHT, H.: Analysis of antibodies against synthetic angiotensin II by the possive haemagglutination technique. Nature (Lond.) **217**, 264–265 (1968).

EDMUNDSON, A. B.: Amino acid sequence of sperm whale myoglobin. Nature (Lond.) **205**, 883–887 (1965).

ELLIOTT, D. F., LEWIS, G. P., HORTON, E. W.: The structure of bradykinin—a plasma kinin from ox blood. Biochem. biophys. Res. Commun. **3**, 87–91 (1960).

ENGLE, J., KURTZ, J., KATCHALSKI, E., BERGER, A.: Polymers of tripeptides as collagen models. II. Conformational changes of poly (L-prolyl-glycyl-L-proline) in solution. J. molec. Biol. **17**, 255–272 (1966).

FEARNEY, F. J., Immunogenicity and specificity studies of peptides and protein conjugates. M. S. Thesis, San Francisco State College, San Francisco, California (1970).

FEARNEY, F. J., LEUNG, C. Y., YOUNG, J. D., BENJAMINI, E.: The specificity of antibodies to a peptide determinant of the tobacco mosaic virus protein induced by immunization with the peptide conjugate. Biochim. biophys. Acta (Amst.) **243**, 509–514 (1971).

FELBER, J. P., ASHCROFT, S. H. J., VILLANUEVA, A., VANNOTTI, A.: Antibodies to synthetic corticotropin. Nature (Lond.) **211**, 654–655 (1966).

FELBER, J. P., MICHELI, A.: Studies on the relationship between antigenicity, the biological activity and the structure of hormones other than insulin. In: Antibodies to biologically active molecules (B. CINADER, ed.), p. 277–300. Oxford: Pergamon Press 1967.

FINN, F. M., HOFMANN, K.: Studies on polypeptides. XXXIII. Enzymic properties of partially synthetic ribonucleases. J. Amer. chem. Soc. **87**, 645–651 (1965).

FISCHER, K., HACHMEISTER, V., KRACHT, J.: Preparation and determination of the antibodies to β^{1-24} corticotropin. Naturwissenschaften **52**, 347–348 (1965).

FLEISCHER, N., GIVENS, J. R., ABE, K., NICHOLSON, W. E.: Studies of ACTH antibodies and their reactions with inactive analogues of ACTH. J. clin. Invest. **44**, 1047–1048 (1965).

FLEISCHER, N., GIVENS, J. R., ABE, K., NICHOLSON, W. E., LIDDLE, G. W.: ACTH antibodies and their reactions with inactive analogs of ACTH. Endocrinology **78**, 1067–1075 (1966).

FUJIO, H., IMANISHI, M., NISHIOKA, K., AMANO, T.: Antigenic structures of hen egg white lysozyme. II. Significance of the N- and C-terminal region as an antigenic site. Biken J. **11**, 207–218 (1968a).

FUJIO, H., IMANISHI, M., NISHIOKA, K., AMANO, T.: Proof of independency of two antigenic sites in egg white lysozyme. Biken J. **11**, 219–223 (1968b).

FUJIO, H., KISHIGUCHI, S., SHINKA, S., SAIKI, Y., AMANO, T.: Immunochemical studies on lysozyme. I. Comparative studies of lysozyme and lysozyme methyl ester. Biken J. **2**, 55–76 (1959).

FUJIO, H., SAIKI, Y., IMANISHI, M., SHINKA, S., AMANO, T.: Immunochemical studies on lysozyme. III. Cross reaction of hen and duck lysozymes and their methyl esters. Biken J. **5**, 201–226 (1962).

GELZER, J.: Immunochemical study of β-corticotropin (1–24)-tetracosa peptide. Immunochemistry **5**, 23–31 (1968).

GERWING, J., THOMPSON, K.: Studies on the antigenic properties of egg white lysozyme. I. Isolation and characterization of a tryptic peptide from reduced and alkylated lysozyme exhibiting haptenic activity. Biochemistry **7**, 3888–3892 (1968).

GOODFRIEND, T. L., FASMAN, G., KEMP, D., LEVINE, L.: Immunochemical studies of angiotensin. Immunochemistry **3**, 223–231 (1966).

GOODFRIEND, T. L., LEVINE, L., FASMAN, G. D.: Antibodies to bradykinin and angiotensin: A use of carbodiimides in immunology. Science **144**, 1344–1346 (1964).

GOODMAN, J. W.: Immunochemical specificity: Recent conceptual advances. Immunochemistry **6**, 139–149 (1969).

GOODMAN, J. W., NITECKI, D. E., STOLTENBERG, I. M.: Immunochemical studies on the poly-γ-D-glutamyl capsule of *Bacillus anthracis*. III. The activity with rabbit antisera of peptides derived from the homologous polypeptide. Biochemistry **7**, 706–710 (1968).

GREGORY, R. A.: Memorial Lecture: The isolation and chemistry of gastrin. Gastroenterology **51**, 953–959 (1966).

GRODSKY, G. M.: Insulin structure as related to biologic and immunologic activity. Rept. Ross. Pediat. Res. Conf. **51**, 8–16 (1965).

GRODSKY, G. M., PENG, C. T., FORSHAM, P. H.: Effect of modification of insulin on specific binding in insulin-resistant sera. Arch. Biochem. **81**, 264–272 (1959).

GROSS, E., WITKOP, B.: Nonenzymic cleavage of peptide bonds: The methionine residues in bovine pancreatic ribonuclease. J. biol. Chem. **237**, 1856–1860 (1962).

HABEEB, A. F. S. A., ATASSI, M. Z.: Enzymic and immunochemical properties of lysozyme. II. Conformation, immunochemistry, and enzymic activity of a derivative modified at tryptophan. Immunochemistry **6**, 555–566 (1969).

HABEEB, A. F. S. A., ATASSI, M. Z.: Enzymic and immunochemical properties of lysozyme. Evaluation of several amino groups by reversible blocking reagents. Biochemistry **9**, 4939–4944 (1970).

HABER, E., PAGE, L. B., JACOBY, G. A.: Synthesis of antigenic branch chain copolymers of angiotensin and poly-L-lysine. Biochemistry **4**, 693–698 (1965).

HABER, E., RICHARDS, F. F., SPRAGG, J., AUSTEN, K. F., VALLOTON, M., PAGE, L. B.: Modification in the heterogeneity of the antibody response. In: Cold Spr. Harb. Symp. quant. Biol. **32**, 299–310 (1967).

HALIKIS, D. N., ARQUILLA, E. R.: Studies on the physical, immunological and biological properties of insulin conjugated with fluorescein isothiocynate. Diabetes **10**, 142–147 (1961).

HARRIS, J. I., KNIGHT, C. A.: Studies on the action of carboxypeptidase on tobacco mosaic virus. J. biol. Chem. **214**, 215–230 (1955).

HEDWALL, P. R.: Effect of rabbit antibodies against angiotensin II on the pressor response to angiotensin II and renal hypertension in the rat. Brit. J. Pharmacol. **34**, 623–629 (1968).

HOLLEMANS, J., VAN DER MEER, J., TOUBER, J. L.: Radioimmunoassay of angiotensin. Nature (Lond.) **217**, 277–278 (1968).

HURWITZ, E., DIETRICH, F. M., SELA, M.: A sensitive technique for detecting and estimating the peptide hormone angiotensin II-β amide and its antibodies by using chemically modified bacteriophage and activated Sepharose. Europ. J. Biochem. **17**, 272–277 (1970).

IMURA, H., SPARKS, L. L., GRODSKY, G. M., FORSHAM, P. H.: Immunologic studies of adrenocorticotropic hormone; dissociation of biologic and immunologic activities. J. clin. Endocr. **25**, 1361–1369 (1965).

ISAGHOLIAN, L. B., BROWN, R. K.: Interaction of a peptide with antibody to oxidized ribonuclease. Immunochemistry **7**, 167–174 (1970).

JASIN, H. E., GLYNN, L. E.: The antigenicity of some synthetic poly-imino acids. I. Antigenic properties of poly L-proline. Immunology **8**, 95–105 (1965a).

JASIN, H. E., GLYNN, L. E.: The antigenic properties of some synthetic poly-imino acids. II. The antigenicity of polypeptides related to collagen; peptides containing hydroxyproline and acetyl-hydroxyproline. Immunology **8**, 260–269 (1965b).

KABAT, E. A.: The nature of an antigenic determinant. J. Immunol. **97**, 1–11 (1966).

KABAT, E. A.: Structural concepts in immunology and immunochemistry. New York: Holt, Rinehart and Winston, Inc. 1968.

KAMINSKI, M.: The analysis of the antigenic structure of protein molecules. In: Progr. Allergy **9**, 79–157 (1965).

KARTHA, G., BELLO, J., HARKER, D.: Tertiary structure of ribonuclease. Nature (Lond.) **213**, 862–865 (1967).

KARUSH, F.: Immunologic specificity and molecular structure. Advanc. Immunol. **2**, 1–40 (1962).

KELLY, B., GERWING LEVY, J.: Immunological studies on the major haptenic peptides from performic acid oxidized ferredoxin from *Clostridium pasteurianum*. Biochemistry **10**, 1763–1766 (1971).

KENDREW, J. C., WATSON, H. C., STRANDBERG, B. E., DICKERSON, R. E., PHILLIPS, D. C., SHORE, V. C.: A partial determination by x-ray methods, and its correlation with chemical data. Nature (Lond.) **190**, 666–670 (1961).

KERP, L., KASEMIR, H., KIELING, F., STEINHILBER, S.: Localization of antibody-binding sites on insulin molecules. Int. Arch. Allergy **36**, 143–151 (1969).

KERP, L., KASEMIR, H., KIELING, F., STEINHILBER, S.: Localization of antibody binding sites on the insulin molecule. Verh. dtsch. Ges. inn. Med. **74**, 552–555 (1968).

KETTMAN, J. R., BENJAMINI, E., MICHAELI, D., LEUNG, D. Y. K.: The synthesis and immunological activity of a peptide related to collagen. Biochem. biophys. Res. Commun. **29**, 623–627 (1967).

KITAGAWA, M., ONOUE, K., OKAMURA, Y., ANAI, M., YAMAMURA, Y.: Immunochemical studies of insulin. II. The specificity of insulin neutralizing antibody and experimental diabetes. J. Biochem. **48**, 483–489 (1960).

Knight, C. A.: Some immunochemical aspects of tobacco mosaic virus. In: Immunochemical approaches to problems in microbiology (Heidelberger, M., O. J. Plescia, and R. A. Day, eds.), p. 161–170. New Brunswick: Rutgers University Press 1961.

Kraus, L. M.: Antigenic determinants in human hemoglobins: The α polypeptide chain. Automation in analytical chemistry. Technicon Symposium 1966, Mediad, N.Y., p. 424–426 (1967a).

Kraus, L. M.: Antigenic determinants in human hemoglobins: The α polypeptide chain. J. Immunol. **99**, 894–898 (1967b).

Kraus, L. M., Sassano, F. G.: Antigenicity of the peptide chains isolated from human hemoglobins A and F. Proc. Soc. exp. Biol. (N.Y.) **118**, 1037–1042 (1955).

Landsteiner, K.: The specificity of serological reactions, 2nd edit. Cambridge, Mass.: Harvard University Press 1945.

Lee, T. H., Lerner, A. B., Buettner-Janusch, V.: On the structure of human corticotropin (adrenocorticotropic hormone). J. biol. Chem. **236**, 2970–2974 (1961).

Leung, C. Y., Wong, K. J., Michaeli, D., Benjamini, E.: The specificity of antibodies induced by immunization with a conjugate of a synthetic peptide related to collagen. In preparation.

Lockwood, D. H., Prout, T. E.: Isoantibodies to insulin. Clin. Res. **10**, 401 (1962).

Margoliash, E., Nisonoff, A., Reichlin, M.: Immunological activity of cytochrome C. I. Precipitating antibodies to monomeric vertebrate cytochromes C. J. biol. Chem. **245**, 931–939 (1970).

Margoliash, E., Reichlin, M., Nisonoff, A.: The relation of immunological activity and primary structure in cytochrome C. In: Conformation of biopolymers (G. N. Ramachandran, ed.), p. 253–277. New York: Academic Press 1967a.

Margoliash, E., Reichlin, M., Nisonoff, A.: The immunological properties of cytochrome C. In: Structure and function of cytochromes. Proc. Symp. Struct. Chem. Aspects of Cytochromes, Osaka, p. 269–280, 1967b.

Margoliash, E., Schejter, A.: Cytochrome C. Advanc. Protein Chem. **21**, 113–286 (1966).

Maron, E., Arnon, R., Sela, M., Perin, J. P., Jolles, P.: Immunological comparisons of bird and human lysozymes and of their "loop" regions. Biochim. biophys. Acta (Amst.) **214**, 222–224 (1970).

Maron, E., Shiozawa, C., Arnon, R., Sela, M.: Chemical and immunological characterization of a unique antigenic region in lysozyme. Biochemistry **10**, 763–771 (1971).

Matsubara, H., Smith, E. L.: The amino acid sequence of human heart cytochrome C. J. biol. Chem. **237**, 3575–3576 (1962).

Maurer, P. H.: Use of synthetic polymers of amino acids to study the basis of antigenicity. Progr. Allergy **8**, 1–40 (1964).

May, J. E., Brown, R. K.: The immunologic role of methionine and cysteine residues in ribonuclease. Immunochemistry **5**, 79–86 (1968).

Merigan, T. C., Potts, J. T.: The essential role of specific conformation in the antigenicity of ribonuclease. Biochemistry **5**, 910–916 (1966).

McGuigan, J. E.: Antibodies to the carboxyl-terminal tetrapeptide of gastrin. Gastroenterology **53**, 697–705 (1967).

McGuigan, J. E.: Immunochemical studies with synthetic human gastrin. Gastroenterology **54**, 1005–1011 (1968a).

McGuigan, J. E.: Antibodies to the C-terminal tetrapeptide amide of gastrin; assessment of antibody binding to cholecystokinin-pancreozymin. Gastroenterology **54**, 1012–1017 (1968b).

McGuigan, J. E.: Studies of the immunochemical specificity of some antibodies to human gastrin I. Gastroenterology **56**, 429–438 (1969a).

McGuigan, J. E.: Binding of caerulein by antibodies to human gastrin I. Gastroenterology **56**, 858–861 (1969b).

Metzger, H., Wofsy, L., Singer, S. J.: A specific antibody-hapten reaction with novel spectral properties. Arch. Biochem. **103**, 206–215 (1963).

Mills, J. A., Haber, E.: The effect on antigenic specificity of changes in the molecular structure of ribonuclease. J. Immunol. **91**, 536–540 (1963).

Moloney, P. J., Coval, M.: Antigenicity of insulin: Diabetes induced by specific antibodies. Biochem. J. **59**, 179–185 (1955).

MUIRHEAD, H., PERUTZ, M. F.: Structure of hemoglobin. A three-dimensional Fourier synthesis of reduced human hemoglobin at 5.5 Å resolution. Nature (Lond.) **199**, 633–638 (1963).

MUTT, V., JORPES, J. E.: Isolation of aspartyl-phenylalanine amide from cholecystokinin-pancreozymin. Biochem. biophys. Res. Commun. **26**, 392–397 (1967).

MITCHELL, B., GERWING LEVY, J., NITZ, R.: Synthesis of the haptenic C-terminal octapeptides of two cross-reacting ferredoxin molecules. Biochemistry **9**, 1837–1844 (1970).

NEUMANN, H., STEINBERG, I. Z., BROWN, J. R., GOLDBERGER, R. F., SELA, M.: On the non-essentiality of two specific disulfide bonds in ribonuclease for its biological activity. Europ. J. Biochem. **3**, 171–182 (1967).

NICOL, D. S. H. W., SMITH, L. F.: Amino acid sequence of human insulin. Nature (Lond.) **187**, 483–485 (1960).

NISONOFF, A., MARGOLIASH, E., REICHLIN, M.: Antibodies to rabbit cytochrome C arising in rabbits. Science **155**, 1273–1275 (1969).

NISONOFF, A., REICHLIN, M., MARGOLIASH, E.: Immunological activity of cytochrome C. II. Localization of a major antigenic determinant of human cytochrome C. J. biol. Chem. **245**, 940–946 (1970).

NITECKI, D. E., GOODMAN, J. E.: Immunochemical studies on the poly-γ-D-glutamyl capsule of *Bacillus anthracis*. II. The synthesis of eight dipeptides and four tripeptides of glutamic acid. Biochemistry **5**, 665–673 (1966).

NITZ, R. M., MITCHELL, B., GERWING, J., CHRISTENSEN, J.: Studies on the antigenicity of bacterial ferredoxins. I. The effect of modification of cysteinyl residues on the antigenicity of *Clostridium pasteurianum* ferredoxin. J. Immunol. **103**, 319–323 (1969).

NOBLE, R. W., REICHLIN, M., GIBSON, Q. H.: The reaction of antibodies to hemeprotein antigens. The measurement of reaction kinetics and stoichiometry by fluorescence quenching. J. biol. Chem. **244**, 2403–2411 (1969).

OKEN, D. E., BIBER, T. V. L.: Biologically effective immunization against angiotensin. Amer. J. Physiol. **214**, 791–795 (1968).

OVARY, Z.: Antigenicity of hemoglobin and its constituents. I. Antigenicity of human adult hemoglobin and its structural units (α and β chains). Immunochemistry **1**, 241–248 (1964).

PATTERSON, R., LUCENA, G., METZ, R., ROBERTS, M.: Reaginic antibody against insulin: Demonstration of antigenic distinction between native and extracted insulin. J. Immunol. **103**, 1061–1071 (1969).

PIEZ, K. A., BLADEN, H. A., LAND, J. M., MILLER, E. J., BORNSTEIN, P., BUTLER, W. T., KANG, A. H.: Comparative studies on the chemistry of collagen utilizing cyanogen bromide cleavage. In: Brookhaven Symposia in Biology, No 21, p. 345–351. Brookhaven, New York: Brookhaven National Laboratory 1968.

PELICHOVA, H., SUZUKI, T., CINADER, B.: Enzyme activation by antibody. II. The distribution of activating antibody in chromatographic fractions of ribonuclease antisera and their competition with inhibiting antibody for enzyme sites. J. Immunol. **104**, 195–202 (1970).

PERLSTEIN, M. T., ATASSI, M. Z.: Accurate delineation of the antigenic reactive region in the sequence 1–55 of myoglobin. Fed. Proc. **30**, 531 (1971).

PERUTZ, M. F.: Structure and function of hemoglobin. I. A tentative atomic model of horse oxyhaemoglobin. J. molec. Biol. **13**, 646–668 (1965).

PERUTZ, M. F., LIQUORI, A. M., EIRICH, F.: X-ray and solubility studies of hemoglobin of sickle-cell anaemia patients. Nature (Lond.) **167**, 929–931 (1951).

POPE, C. G.: The immunology of insulin. Advanc. Immunol. **5**, 209–244 (1966).

PRESSMAN, D., GROSSBERG, A. L.: The structural basis of antibody specificity. New York: W. A. Benjamin Inc. 1968.

PROUT, T. E.: The antigenicity of insulin: A review. J. chron. Dis. **15**, 879–885 (1962).

RAPPAPORT, I.: The antigenic structure of tobacco mosaic virus. In: Advanc. Virus Res. **2**, 223–275 (1965).

RAPPAPORT, I., ZAITLIN, M.: Conformational changes in the antigenic determinant of tobacco mosaic virus protein resulting from polymerization of the subunits. Virology **41**, 208–217 (1970).

REICHLIN, M.: Effects on antigenicity of single amino acid changes in human hemoglobin. Fed. Proc. **28**, 435 (1969).

REICHLIN, M.: The distribution of specificity in rabbit antisera directed toward human hemoglobins. Immunochemistry **7**, 15–28 (1970).

REICHLIN, M., BUCCI, E., ANTONI, E., WYMAN, J., ROSSI-FANELLI, A.: The immunochemical difference between horse oxy- and deoxyhemoglobin. J. molec. Biol. **9**, 785–788 (1964a).

REICHLIN, M., BUCCI, E., FRONTICELLI, C., WYMAN, J., ANTONINI, E., IOPPOLO, C., ROSSI-FANELLI, A.: The properties and interactions of isolated α and β chains of human hemoglobins. IV. Immunological studies involving antibodies against the isolated chains. J. molec. Biol. **17**, 18–28 (1966).

REICHLIN, M., BUCCI, E., FRONTICELLI, C., WYMAN, J., ANTONINI, E., ROSSI-FANELLI, A.: Properties and interactions of the isolated α and β chains of human hemoglobin. II. Immunochemical behaviour of the chains: Fractionation of an antigen. J. molec. Biol. **12**, 774–779 (1965a).

REICHLIN, M., BUCCI, E., WYMAN, J., ANTONINI, E., ROSSI-FANELLI, A.: The role of α and β chains in the immunochemical difference between oxy and deoxy human hemoglobin. J. molec. Biol. **11**, 775–784 (1965b).

REICHLIN, M., HAY, M., LEVINE, L.: Immunochemical studies of hemoglobin and myoglobin and their globin moieties. Biochemistry **2**, 971–979 (1963).

REICHLIN, M., HAY, M., LEVINE, L.: Antibodies to human A_1 hemoglobin and their reaction with A_2 S, C., and H hemoglobins. Immunochemistry **1**, 21–30 (1964b).

REICHLIN, M., MARGOLIASH, E., NISONOFF, A.: Immunogenicity of polymers of cytochrome C. Fed. Proc. **27**, 276 (1968a).

REICHLIN, M., SCHNURE, J. J., VANCE, V. K.: Induction of antibodies to porcine ACTH in rabbits with nonsteroidogenic polymers of BSA and ACTH. Proc. Soc. exp. Biol. (N.Y.) **128**, 347–350 (1968b).

RICHARDS, F. M., VITHAYATHIL, P. J.: The preparation of subtilisin-modified ribonuclease and the separation of the peptide and protein components. J. biol. Chem. **234**, 1459–1465 (1959).

ROELANTS, G. E., GOODMAN, J. W.: Tolerance induction by an apparently non-immunogenic molecule. Nature (Lond.) **227**, 175–176 (1970).

ROYER, G., CANADY, W. J.: A common size parameter for hydrophobic binding of inhibitors by α-chymotrypsin, alkylated α-chymotrypsin and yeast-alcohol-dehydrogenase. Arch. Biochem. **124**, 530–534 (1968).

SALVIN, S. B., LIAUW, H. L.: Hypersensitivity to peptide fragments. Int. Arch. Allergy **31**, 366–379 (1967).

SCHECHTER, I., SCHECHTER, B., SELA, M.: Combining sites of antibodies with L-alanine and D-alanine peptide specificity and the effect of serum proteolytic activity on their activation. Biochim. biophys. Acta (Amst.) **127**, 438–456 (1966).

SCHERAGA, H. A.: Structural studies of pancreatic ribonuclease. Fed. Proc. **26**, 1380–1387 (1967).

SCHWICK, H. G.: Zur Immunologie des Insulins eine Literaturübersicht. Behringwerk-Mitt. **46**, 87–106 (1966).

SELA, M.: Immunological studies with synthetic polypeptides. Advanc. Immunol. **5**, 29–129 (1966).

SELA, M.: Use of synthetic polypeptides in studies on the control of antibody formation. Nobel Symposium 3. Gamma globulins, p. 455–471 (J. KILLANDER, ed.). Stockholm: Interscience Pub. 1967.

SELA, M.: Antigenicity: Some molecular aspects. Science **166**, 1365–1374 (1969).

SELA, M., ANFINSEN, C. B., HARRINGTON, W. F.: The correlation of ribonuclease activity with specific aspects of tertiary structure. Biochim. biophys. Acta (Amst.) **26**, 502–512 (1957).

SELA, M., ARNON, R.: Studies on the chemical basis of the antigenicity of proteins. 3. The role of rigidity in the antigenicity of polypeptidyl gelatins. Biochem. J. **77**, 394–399 (1960).

SELA, M., SCHECHTER, B., SCHECHTER, I., BOREK, F.: Antibodies to sequential and conformational determinants. In: Cold Spr. Harb. Symp. quant. Biol. **32**, 537–545 (1967).

SENGBUSCH, P. VON: Aminosäureaustausch und Tertiarstruktur eines Proteins. Vergleich von Mutanten des Tabakmosaikvirus mit serologischen und physikochemischen Methoden. Z. Vererbungsl. **96**, 364–386 (1965).

SHINKA, S., IMANISHI, M., KUWAHARA, O., AMANO, T.: Immunochemical studies on lysozyme. II. On the non-neutralizing antibodies. Biken J. **5**, 181–200 (1962).

SHINKA, S., IMANISHI, M., MIYAGAWA, N., AMANO, T., INOUYE, M., TSUGITA, A.: Chemical studies on antigenic determinants of hen egg white lysozyme. I. Biken J. **10**, 89–107 (1967a).

SHINKA, S., IMANISHI, M., MIYAGAWA, N., AMANO, T., INOUYE, M., TSUGITA, A.: Antigenic determinants of hen egg white lysozyme. Proc. Jap. Acad. **43**, 505–510 (1967b).

SINGER, S. J.: Structure and function of antigen and antibody proteins. In: The proteins, 2nd edit., vol. III (H. NEURATH, ed.), p. 269–357. New York: Academic Press 1965.

SINGER, S. J., RICHARDS, F. M.: The interaction of sublitisin-modified ribonuclease with antisera against native ribonuclease. J. biol. Chem. **234**, 2911–2914 (1959).

SKEGGS, L. T., JR., KAHN, J. R., LENTZ, K., SHUMWAY, N. P.: The preparation, purification, and amino acid sequence of a polypeptide renin substrate. J. exp. Med. **106**, 439–453 (1957).

SMYTH, D. G., STEIN, W. H., MOORE, S.: The sequence of amino acid residues in bovine pancreatic ribonuclease: Revisions and confirmations. J. biol. Chem. **238**, 227–234 (1963).

SPITLER, L., BENJAMINI, E., YOUNG, J. D., KAPLAN, H., FUDENBERG, H. H.: Studies on the immune response to a characterized antigenic determinant of tobacco mosaic virus protein. J. exp. Med. **131**, 133–148 (1970).

SPRAGG, J., AUSTEN, K. F., HABER, E.: Production of antibody against bradykinin: demonstration of specificity by complement-fixation and radio-immunoassy. J. Immunol. **96**, 865–871 (1966).

SPRAGG, J., SCHROEDER, E., STEWART, J. M., AUSTEN, K. F., HABER, E.: Structural requirement for binding to antibody of sequence variants of bradykinin. Biochemistry **6**, 3933–3941 (1967).

SPRAGG, J., TALAMO, R. C., SUZUKI, K., APPELBAUM, D. M., AUSTEN, K. F.: Structural requirements for binding of bradykinin to antibody. II. Studies with bradykinin fragments. Biochemistry **7**, 4086–4089 (1968).

STAVITSKY, A. B.: Micromethod for the study of proteins and antibodies. I. Procedure and general applications of hemagglutination and hemagglutination inhibition reactions with tannic acid and protein-treated red blood cells. J. Immunol. **72**, 360–367 (1954).

STEWART, J. M., YOUNG, J. D., BENJAMINI, E., SHIMIZU, M., LEUNG, C. Y.: Immunochemical studies on the tobacco mosaic virus protein. IV. The automated solid phase synthesis of a decapeptide of tobacco mosaic virus protein and its reaction with antibodies to the whole protein. Biochemistry **5**, 3396–3400 (1966).

SUZUKI, T., PELICHOVA, H., CINADER, B.; Enzyme activation by antibody. I. Fractionation of immune sera in search for an enzyme activating antibody. J. Immunol. **103**, 1366–1376 (1969).

TANAKA, M., NAKASHIMA, T., BENSON, A., MOWER, H., YASUNOBU, K. T.: The amino acid sequence of *Clostridium pasteurianum* ferredoxin. Biochemistry **5**, 1666–1680 (1966).

TANFORD, C., HAUENSTEIN, J. D.: Identification of the chemical difference between chromatographic components of ribonuclease. Biochim. biophys. Acta (Amst.) **19**, 535–539 (1965).

TOUBER, J. L., STOLL, R. W., ENSINCK, J. W., WILLIAMS, R. H.: Immunological studies of the A and B chains of insulin. Diabetes **119**, 409–419 (1970).

TRAUB, W., YONATH, A.: Polymers of tripeptides as collagen models. I. X-ray studies of poly (L-prolyl-glycyl-L-proline) and related polytripeptides. J. molec. Biol. **16**, 404–414 (1966).

TSUGITA, A., GISH, D., YOUNG, J., FRAENKEL-CONRAT, H., KNIGHT, C. A., STANLEY, W. M.: The complete amino acid sequence of the protein of tobacco mosaic virus. Proc. nat. Acad. Sci. (Wash.) **46**, 1463–1469 (1960).

VAN REGENMORTEL, M. H. V.: Plant virus serology. Advanc. Virus Res. **12**, 207–271 (1966).

VAN VUNAKIS, H., LEIKHIM, E., DELANEY, R., LEVINE, L., BROWN, R. K.: Studies on the antigenic structure of ribonuclease. II. Partially deaminated ribonuclease. J. biol. Chem. **235**, 3430–3436 (1960).

Varandini, P. T.: Studies on the nature of antigenicity of A and B chains of bovine insulin. Biochemistry **6**, 100–104 (1967).
Williams, A. W., Reichlin, M.: Stoichiometry of an antigen-antibody reaction. Fed. Proc. **27**, 259 (1968).
Wilson, S.: Immunology of protein hormones. In: Antibodies to biologically active molecule (B. Cinader, ed.), p. 235–276. Oxford: Pergamon Press 1967.
Wilson, S., Aprile, M. A., Sasaki, L.: The antigenic loci of insulin. Canad. J. Biochem. **45**, 1135–1144 (1967).
Wyckoff, H. W., Hardman, K. D., Allewell, N. M., Inagami, T., Johnson, L. M., Richards, F. M.: The structure of ribonuclease-S at 3.5 Å resolution. J. biol. Chem. **242**, 3984–3988 (1967).
Yagi, Y., Maier, P., Pressman, D.: Antibodies against the component polypeptide chains of bovine insulin. Science **147**, 617–619 (1965).
Yagi, Y., Maier, P., Pressman, D.: Antigenic determinant regions in the phenylalanyl chain of bovine insulin. Fed. Proc. **26**, 703 (1967).
Young, J. D., Benjamini, E., Leung, C. Y.: Immunochemical studies on the tobacco mosaic virus protein. VIII. Solid phase synthesis and immunological activity of peptide related to an antigenic area of tobacco mosaic virus protein. Biochemistry **7**, 3113–3119 (1968).
Young, J. D., Benjamini, E., Shimizu, M., Leung, C. Y.: Immunochemical studies on the tobacco mosaic virus protein. III. The degradation of an immunologically active tryptic peptide of tobacco mosaic virus protein and the reactivity of the degradation products with antibodies to the whole protein. Biochemistry **5**, 1481–1488 (1966).
Young, J. D., Benjamini, E., Stewart, J. M., Leung, C. Y.: Immunochemical studies on tobacco mosaic virus protein. V. Solid phase synthesis of peptides of an antigenically active decapeptide of tobacco mosaic virus protein and the reaction of these peptides with antibodies to the whole protein. Biochemistry **6**, 1455–1460 (1967).
Young, J. D., Leung, C. Y.: Immunochemical studies on lysozyme and carboxvmethylated lysozyme. Biochemistry **9**, 2755–2762 (1970).

Replication and Lysogeny with Phage P22 in *Salmonella typhimurium*

MYRON LEVINE[1]

With 2 Figures

Contents

I. Introduction

The bacterial viruses have a long history of usefulness in the analysis of fundamental genetic problems. Vegetative replication of temperate phage requires the controlled and orderly expression of most of the phage genes. An intricate regulatory system must be evoked for the establishment of lysogeny. Virtual shut-down of all vegetative phage functions takes place. A small number of phage genes remain active to carry out the reactions necessary to accomplish integration of the phage genome into that of the host. Only a few phage genes are expressed in the lysogen. The genes required for vegetative replication can be reactivated on induction. This, in effect, constitutes a differentiated system. Not only are the regulatory mechanisms operating in lysogeny important in their own right, but they are also valuable models for understanding the emerging problems of oncogenic virus-host interactions.

Phage P22 is a temperate phage of *Salmonella typhimurium* (ZINDER and LEDERBERG, 1952; ZINDER, 1953). On infection, some bacteria lyse, giving rise to a yield of progeny particles, and some survive, giving rise to lysogenic cells.

[1] Department of Human Genetics, University of Michigan, Ann Arbor, Michigan 48104.

Lysogenic bacteria are immune to superinfection by the phage, but they possess the ability to produce the originally infecting phage on *induction.* Genetic control of the lysogenic condition lies in the *prophage,* which is the phage hereditary material integrated into the bacterial genome.

The establishment of the lysogenic state occurs in two stages. The first is the repression of vegetative phage development and survival of the infected cell, and the second is the physical incorporation of prophage into the bacterial chromosome. Failure to initiate and maintain repression of phage vegetative functions results in progeny formation and cell lysis: the *lytic* or *productive response* (LWOFF, 1953). Prophage integration does not automatically follow from successful repression of phage growth. A number of additional steps are needed for incorporation of prophage. We will adopt the term *reduction* (LWOFF, 1953) for the initial repression of phage development and *integration* for prophage incorporation. The complete process is called the *lysogenic response.*

This review will describe the physical and genetic properties of phage P22, the reactions in the lytic and lysogenic responses, and the control mechanisms as they are now understood. In many of its interactions with its host bacterium, the phage P22 presents unique features for temperate phage systems.

II. The Phage Particle

Phage P22 is structurally a relatively simple representative of the larger phages, consisting of a polyhedral head, about 600 Å in diameter, to one apex of which is attached a small base plate (YAMAMOTO and ANDERSON, 1961; ISRAEL et al., 1967). The base plate consists of a central core around which are arranged six spikes. In the absence of the spikes, a small neck-like structure is visible (ISRAEL et al., 1967). Packaged into the head of each phage is a double-stranded linear DNA molecule of 26 to 27 million molecular weight and a mean contour length of 13.17 μ (RHOADES et al., 1968). The density of the phage DNA is 1.7069 and that of the LT2 host strain is 1.7106 (SMITH, 1968). P22 DNA is terminally redundant but does not have complementary single-stranded cohesive ends (RHOADES et al., 1968) as do some other temperate phage DNA molecules. Digestion with exdonuclease III produces molecules which circularize upon annealing. This indicates a duplex terminal repetition of base sequence in P22 DNA. The length of the repeated region has been estimated to comprise 2.5 to 5% of the molecule, or 1000–2000 base pairs. Circular molecules also can be produced by denaturing and reannealing P22 DNA, suggesting that the DNA molecules released by P22 particles are circular permutations of one another. This is supported by competition-annealing experiments which indicate that short fragments derived from the ends of P22 DNA molecules contain all of the sequences present in a complete molecule. The number of cyclic permutations present in a population of phage particles is large, possibly equal to the number of base pairs in the genome. In its physical characteristics, P22 DNA resembles more closely molecules of the virulent phages T2 and T4 than those of the temperate phages λ, P2 and their relatives.

III. Vegetative and Prophage Linkage Maps

A large number of conditional lethal mutations of both temperature sensitive (*ts*) (LEVINE, BOTSTEIN and colleagues) and suppressor sensitive (amber and UGA) (LEVINE, BOTSTEIN and colleagues; KOLSTAD and PRELL, 1969; LEW and ROTH, 1970) types have been isolated. They have been arranged by complementation studies into some 25 cistrons. Together with plaque-morphology mutations and mutations affecting lysogeny, at least 30 loci have been defined on P22. Phage P22 DNA can probably code for about 50 genes.

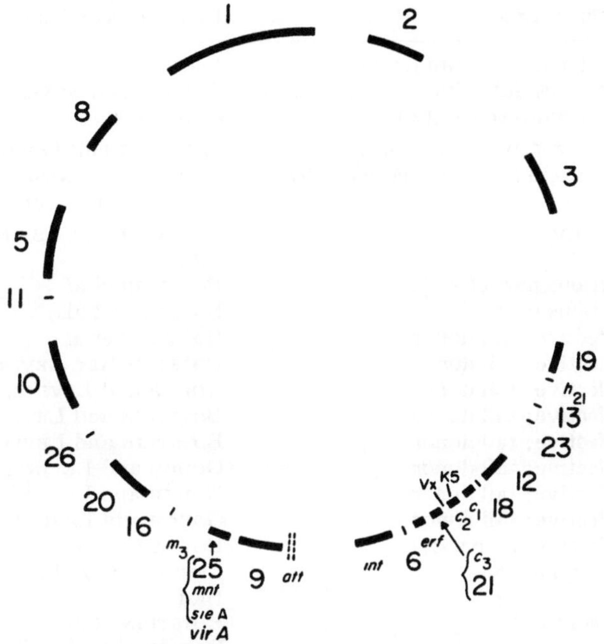

Fig. 1. Vegetative linkage map of phage P22. Based on data presented by GOUGH and LEVINE (1968) and CHAN and BOTSTEIN (unpublished) and compiled by D. BOTSTEIN. The bars represent the recombination frequencies between the most distant mutations in each gene

As might be expected from the circularly permuted nature of P22 DNA, vegetative recombination studies place these loci on a circular linkage map (GOUGH and LEVINE, 1967; KOLSTAD and PRELL, 1969). The order and relationship of loci to one another (Fig. 1) has been deduced from two and three factor crosses and confirmed by complementation studies with prophage deletion mutants (CHAN and BOTSTEIN, unpublished). The prophage linkage map is a unique linear order of the circular vegetative map as though the circle was opened between gene *9* and the *int* gene and inserted into the host chromosome between the *pro* A and *pro* C loci (Fig. 2) (SMITH and LEVINE, 1965b; SMITH, 1968; CHAN and BOTSTEIN, unpublished). Table 1 lists the various P22 genes in order as they lie on the linkage map and gives a brief description of their mutant phenotypes. They will be described in greater detail in the body of the paper.

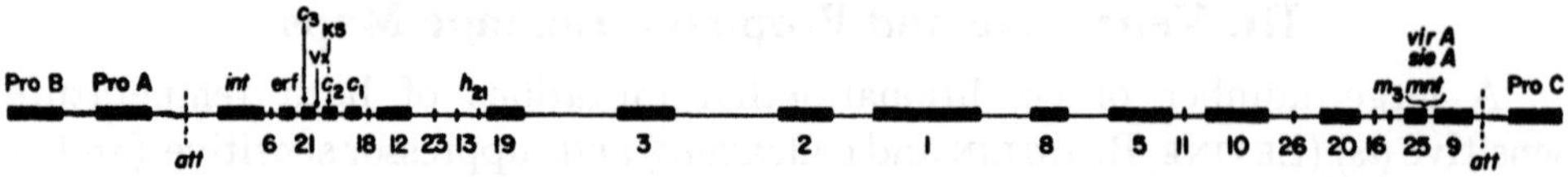

Fig. 2. Prophage map of phage P22. Based on data supplied by SMITH and LEVINE (1965), SMITH (1968) and CHAN and BOTSTEIN (unpublished) and compiled by D. BOTSTEIN. The bars represent the recombination frequencies between the most distant mutations in each gene

Table 1. The Genes of Phage P22

Gene	Mutant Phenotype	References or Source
9	Tail-part defective; head donor	ISRAEL
25	DNA synthesis defective; no association of DNA with intermediate I	BEZDEK and SOSKA, 1970; LEVINE et al., 1970
mnt[a]	Lysogeny defective; maintenance	ZINDER, 1958; GOUGH, 1968
sieA[a]	Failure to exclude superinfecting DNA	RAO, 1968; WALSH and MEYNELL, 1967; SUSSKIND et al., 1971
m3	Plaque morphology	LEVINE, 1957; LEVINE and CURTISS, 1961
16	Noninfectious particles	BOTSTEIN et al.
20	Noninfectious particles	BOTSTEIN et al.
26	Head defective; tail donor	BOTSTEIN et al.
10	Head defective; tail donor	BOTSTEIN and LEVINE, 1968
11	Head defective; tail donor	GOUGH and LEVINE, 1968
5	Head defective; tail donor	BOTSTEIN and LEVINE, 1968
8	Head defective; tail donor	BOTSTEIN and LEVINE, 1968
1	Head defective; tail donor	GOUGH and LEVINE, 1968
2	Head defective; tail donor	GOUGH and LEVINE, 1968
3	Head defective; tail donor	GOUGH and LEVINE, 1968
19	Lysis defective; no lysozyme	BOTSTEIN et al.
h_{21}	Plaque morphology	LEVINE, 1957; LEVINE and CURTISS, 1961
13	Lysis defective	BOTSTEIN et al.
23		BOTSTEIN et al.
12	DNA synthesis defective; integration defective	GOUGH and LEVINE, 1968; LEVINE and SCHOTT, 1971
18	DNA synthesis defective; integration defective	LEVINE and SCHOTT, 1971
c_1	Clear plaques; reduction defective	LEVINE, 1957; LEVINE and CURTISS, 1961
c_2	Clear plaques; reduction defective and maintenance defective	LEVINE, 1957; LEVINE and CURTISS, 1961
c_3[a]	Clear plaques; reduction defective	LEVINE, 1957; LEVINE and CURTISS, 1961
21[a]	Possible DNA synthesis defective	LEVINE, unpublished
erf	Recombination and replication defective, integration defective	YAMAGAMI and YAMAMOTO, 1970; BOTSTEIN and MATZ, 1970
6	Possible DNA synthesis defective[b]	BEZDEK and SOSKA, 1970; LEVINE, unpublished
int	Integration defective	SMITH and LEVINE, 1967; SMITH, 1968

[a] Brackets indicate that the order of these genes is not known.

[b] The *ts 6.1* mutant used by BEZDEK and SOSKA (1970) appears to be different from the mutant of the same name used in my Laboratory. The BEZDEK and SOSKA mutant makes phage DNA at the nonpermissive temperature and has the characteristics of a late mutant. Our mutant is leaky, but appears to be an early gene.

IV. The Lytic Response

Infections of sensitive cells with low multiplicities of wild-type, *P22c*$^+$, favor the lytic response. As the number of phage adsorbed per sensitive cell increases, the frequency of the lytic response decreases and lysogenization increases (Boyd, 1951). One-hundred-percent lytic responses are achieved by infections with the clear mutants, *P22*c_1 or *P22*c_2, which are defective in reduction [(Levine, 1957), see section V a for a more complete description of the clear mutants].

The broad outlines of phage P22 replication and morphogenesis in lytic infection and after induction are discernible. The central role in phage DNA replication is played by intermediate I, one of the intermediates in the synthesis and maturation of phage P22 DNA (Botstein, 1968; Botstein and Levine, 1968a and b). If an infection is initiated with ^{32}P-labeled phage, label is recovered after gentle lysozyme-detergent lysis at the bottom of a sucrose gradient in a position indicating a sedimentation rate of at least 1 000 S. During infection or induction, the radioactivity incorporated into viral DNA during short pulses with ^{3}H-thymidine co-sediments with the ^{32}P-labeled parental DNA. This rapidly sedimenting fraction is called intermediate I. With time both labels are released from intermediate I and are converted into the uniform lengths of DNA in progeny phage particles through a second intermediate form, intermediate II. Intermediate II is a collection of concatemers; covalently bonded multiple phage genomes which sediment in sucrose gradients 1.2–1.5 times as fast as mature phage DNA molecules (Botstein, 1968; Botstein and Levine, 1968a and b).

Intermediate I undoubtedly represents a complex of phage DNA with host cell components. Treatment with pronase, alkali or high salt disrupts the complex with release of DNA. The complex has several properties expected of the replication machinery of phage P22. Intermediate I contains parental DNA, it is the site of synthesis of new DNA, and the DNA associated with it is precursor to the DNA in progeny particles.

Little parental DNA appears in intermediate I before the fifth minute of the infection (Levine et al., 1970). This finding correlates very well with the time of onset of phage DNA synthesis (Smith and Levine, 1965a). Parental DNA accumulates in intermediate I, reaching a peak at about 15 minutes and then decreases. The product of gene *25* seems to be needed very early in the infection for the appearance of parental DNA in the replication complex. Under nonpermissive conditions virtually no *P22 ts25* mutant DNA associates with intermediate I, no phage DNA and protein synthesis takes place and the infected cells survive (Levine et al., 1970; Bezdek and Soska, 1970). In fact, no known phage functions are initiated under these conditions (Levine, Hoffman, Israel, Bronson, Woodworth-Gutai, unpublished; Bezdek and Soska, 1970). Gene *25* mutants are not defective in injection (Hoffman, unpublished). Bezdek and Soska (1970) suggest that gene *25* plays a central regulatory role in P22 replication similar to that of gene *N* for phage λ. Continuous *N* product production is needed for λ production (RABOVSKY and KONRAD, 1970). However, temperature shift experiments with *P22 ts 25* mutants demonstrate that gene *25* product need act for only a very few minutes at the onset of infection for phage production (LEVINE, unpublished). In addition, induction at high temperature of cells lysogenic for *ts25* prophage, results in cell lysis and release of phage (HOFFMAN, unpublished). Gene *25* function can thus be bypassed on induction. Whether gene *25* acts in some direct manner in relation to intermediate I or at some earlier step remains unresolved.

The products of genes *18* and *12*, which map to the right of the *C* region (Fig. 1) are required early in infection for phage DNA synthesis (BEZDEK and SOSKA, 1970; LEVINE and SCHOTT, 1971; LOVEDAY and BOTSTEIN, personal communication). In nonpermissive conditions, *P22ts18* and *P22ts12* fail to undergo even one round of replication in single infection, but complement for DNA synthesis and phage production in mixed infection with each other. In mixed infections with mutant and nonmutant phages, phage DNA synthesis is normal and both phages appear in the yields. That is, the *18* and *12* mutations are recessive to wild-type for phage DNA synthesis.

Clearly, synthesis of phage DNA is not a prerequisite for association of parental DNA with the complex, since the DNA of mutants in genes *18* and *12* associate with intermediate I at restrictive conditions. These findings suggest that gene *12* and gene *18* products are expressed in intermediate I.

Preliminary data suggests that genes *6* and *21*, which map to the left of the *C* region (Fig. 1), may also play a role in phage DNA synthesis. However, the available mutants are leaky, making analyses difficult. The *ts6* mutant used by BEZDEK and SOSKA (1970) appears to be different from our *ts6* mutant, since their mutant makes phage DNA at the nonpermissive temperature and has the characteristics of a late mutant.

In addition to gene *18* and gene *12* products, sustained synthesis of phage DNA appears to require a recombinational event. Mutants of phage P22 have been isolated (YAMAGAMI and YAMAMOTO, 1970; BOTSTEIN and MATZ, 1970) which cannot grow on infection of *recA* mutant recombination deficient host cells (WING et al., 1968). The mutants, designated *erf*$^-$ for *e*ssential *r*ecombination *f*unction, are defective in a phage-specified general recombination system, as shown by two and three factor crosses carried out in *rec*$^+$ and *rec*$^-$ hosts. The host recombination system can be utilized in place of the phage system and, in addition, *erf*$^+$ is dominant to *erf*$^-$. The *erf* gene is located in the "early" region of the phage linkage map (Fig. 1). DNA synthesis is initiated after infection of *rec*$^-$ cells with *P22 erf*$^-$, but proceeds for only a short time, long enough to undergo about one round of replication. BOTSTEIN and MATZ (1970) suggested that DNA synthesis cannot be sustained without circularization of template DNA, presumably by recombination in the terminal redundancies and that this recombinational event does not occur in a *P22 erf*$^-$ infected *rec*$^-$ cell. The abortive replication of DNA might represent synthesis from a unique point of origin in one direction to the ends of the circularly permuted linear molecules. Support for this interpretation will become evident from the description of the effect of the *erf* mutation on lysogenization and induction (section 5). *P22 erf⁻* prophage, on induction, grows normally in *rec⁻* cells, presumably because the prophage genome is excised in a circular form. Late phage proteins are made during *P22 erf⁻* infections of *rec⁻* cells, indicating that sustained DNA synthesis is not required for expression of late P22 functions.

A new nuclease activity specific for native DNA has been demonstrated in $P22c_1$ infected cells after subjecting extracts to DNA-cellulose and DEAE-cellulose chromatography (WOODWORTH-GUTAI, ISRAEL and LEVINE, 1971 and in press). This enzyme activity differs from a similar host activity in Mg^{++} and Mn^{++} optima, pH optimum and heat lability. This new activity is not found following infection of a lysogen with $P22c_1$ or after infection of a nonpermissive host with *P22 amberf⁻* and thus appears to be specified by the phage genome.

Infections under nonpermissive conditions with mutants in genes *5, 8, 1, 2* and *3* show normal accumulation of parental and newly synthesized phage DNA in intermediate I. However, both types of DNA persist in intermediate I in these infections. Little or no DNA is released and no phage lengths of DNA are formed (BOTSTEIN, unpublished). If, however, intermediate I is disrupted, the released DNA bands in sucrose gradients as concatemers. These concatemer molecules may be as large as ten phage genome equivalents. Lysates from infections with all five mutants act as tail donors in the *in vitro* tail-head reconstitution assay for active phage (ISRAEL et al., 1967; ISRAEL, 1967). All of these are head mutants. Gene *5* product appears to be the major protein of the head membrane (BOTSTEIN and KING, unpublished; ROSEN and LEVINE, unpublished). No headlike structures are seen in gene *5* mutant lysates. Lysates of infections under nonpermissive conditions with mutants in gene *8* show an aberrant structure containing gene *5* product (BOTSTEIN and KING, unpublished). Mutant gene 1, 2 and 3 lysates contain head membranes without DNA (BOTSTEIN and KING, unpublished).

Infections under restrictive conditions with mutants in genes *10, 26, 16, 20, 9, 19* and *13* also give normal association of DNA with intermediate I and normal rates of DNA synthesis. DNA is released from intermediate I in these infections and phage lengths of DNA are produced (BOTSTEIN, unpublished). The lysates from infections with mutants in genes *10* and *26* show head membranes. The data suggest that these membranes were filled with phage DNA, but these structures are unstable on lysis. Mutant *9* lysates are the only ones which act as head donors in the *in vitro* reaction. Gene *9* may be the structural gene for the protein of the tail spike (ROSEN and LEVINE, unpublished). Infections with mutants in genes *16* and *20* give rise to encapsulated normal looking phage particles, which are however largely non-infectious (BOTSTEIN, unpublished). The products of genes *19* and *13* are needed for lysis of the cell and release of phage. Gene *19* is the structural gene for a phage induced lysozyme (BOTSTEIN and WADDELL, unpublished). Active phages are produced in these infections under nonpermissive conditions.

COHEN (1969a und b) has described a mutant (Lys^-) which shows delayed lysis. In unsupplemented minimal medium, Lys^- infected cells require more than three hours to lyse. Intracellular phage development and lysozyme synthesis proceeds in Lys^- infections at least as rapidly as in Lys^+ infected cells. Lysis can be induced by addition of cysteine or histidine to the growth medium. The relationship of Lys^- mutants to genes *19* and *13* has not been explored.

The above data implicate the formation of the normal phage head in the release of phage DNA from intermediate I and in the maturation of P22 DNA and support a "headful" model for P22 similar to that proposed by STREISINGER et al. (1964) to explain the circularity of the T4 linkage map. They proposed the formation of long concatemers of phage DNA during the vegetative cycle. As maturation progressed, these would be subdivided into segments of constant length, somewhat larger than a complete genome. This would produce molecules of a standard length but with different ends and would account for the terminal redundancy and circular permutation of the DNA. The length of the chromosome of a mature phage particle might be determined by the amount of DNA that can be contained in the volume of a phage head, a "headful" of DNA. The head membrane might be the measuring and cutting device by which this is achieved. The normal products of head genes may be necessary parts of this device.

Phage DNA in the form of concatemers is formed in intermediate I. The mechanism of formation of concatemers is not established. Dimers and longer concatemers can be produced by recombination between two identical molecules at the terminal redundancies (STREISINGER et al., 1964). Another possibility is that long molecules are produced by the rolling circle mechanism proposed by GILBERT and DRESSLER (1968). Recombination in the terminal redundancies of an individual molecule could give a circular template on which DNA can be replicated continuously around the circle. The need for a circular template is supported by the growth of erf^- prophage after induction of a rec^- lysogen. As suggested above, the DNA is circularized by the prophage excision mechanism. Presumably, concatemers are formed in these induced lysogens. These concatemers would have to be made in the absence of both the host and *erf* recombination mechanisms suggesting that they could not be formed by recombination between different molecules.

V. The Lysogenic Response

A. Reduction

Reduction is an efficient process in P22 infections. At multiplicities of 10 or more wild-type, c^+, phage, over 95 percent of the infected cells survive and give rise to lysogenic progeny. Reduction is under the control of the *C* region of phage P22, comprised of three closely linked genes, c_1, c_2 and c_3 (LEVINE,

1957; LEVINE and CURTISS, 1961). Mutations in any of these genes give rise to phages which produce clear plaques in contrast to the turbid plaques of *P22c*$^+$. *P22*c_1 and *P22*c_2 can give the lytic response only; *P22*c_3 shows decreased levels of lysogenization. Mixed infections between the various mutants and between the mutants and *P22c*$^+$ give normal frequencies of lysogenization.

Insight into the functions of the *C* region comes from the study of the specific patterns of the rates of DNA synthesis in cells infected with *P22c*$^+$ or the clear mutants c_1 or c_2 (SMITH and LEVINE, 1964). Following infection of a log-phase culture there is an immediate depression in the rate of DNA synthesis. At about 4 minutes, a sharp rise in the rate of synthesis in *P22c*$^+$ infected cells occurs and reaches a peak at about six minutes. This burst of synthesis is short lived. A marked depression in rate of DNA synthesis occurs until the sixteenth minute, when the rate rises sharply again, reaching rates similar to those of uninfected cells by 40 minutes. At this point the cells begin to divide. Analysis of the type of DNA synthesized (SMITH and LEVINE, 1965) shows that the six minute peak consists primarily of newly synthesized phage DNA. Bacterial DNA synthesis is progressively inhibited from the beginning of the infection until 16 minutes and then recovers. At the time of recovery of host DNA synthesis, vegetative phage DNA production is completely repressed.

Two sequential functions controlled by genes c_1 and c_2 are required for the repression of phage replication (LEVINE and SMITH, 1964). Failure of either function, as occurs in infections with either *P22*c_1 or *P22*c_2, results in the lytic response and phage production. The ability to repress phage DNA synthesis at six minutes is controlled by the c_1 locus. Temperature pulse experiments with *P22ts* c_1 demonstrate that the c_1 gene product must act during the 6th to 10th minute to initiate the repression. In its absence the rate of phage DNA synthesis increases until 25 minutes when lysis and phage liberation begins (LEVINE and SMITH, 1964; SMITH and LEVINE, 1964). On the other hand, *P22*c_2 infections show the six minute peak in the rate of phage DNA synthesis, with repression setting in as it does in *P22c*$^+$ infections, and lasting until 16 minutes. At this point massive synthesis of phage DNA starts and continues until the end of the latent period at 50 minutes when, again, the cells start to lyse and progeny phage are liberated. In mixed infections with *P22*c_1 and *P22*c_2, patterns of DNA synthesis and lysogenization similar to those of *P22c*$^+$ alone are observed. These lysogens carry *P22*c_1 prophages, which are stable. In contrast, the *P22*c_2 have not been observed as prophage. Therefore, the c_1 gene product is needed for reduction and the establishment of lysogeny, but is not needed for maintenance of the lysogenic state. Temperature shift experiments with *P22 ts* c_2, which exhibit the c^+ phenotype at low temperature and the c_2 phenotype at high temperature, demonstrate that the c_2 gene product must remain active from 16 minutes on to maintain the repression on phage DNA synthesis (LEVINE and SMITH, 1964). This product appears to be required early for reduction and then continuously for maintenance of lysogeny (see section Vc). The function of the c_3 gene is not now understood, except that it *is* required for establishment but not maintenance of the lysogenic state. Once established, a c_3 mutant prophage is as stable as wild-type.

The physiological state of the host cell and the conditions of culture are known to influence the lysogenic response. HONG et al. (1971) have reported a striking example of this. Mutants of *Salmonella typhimurium* defective in adenyl cyclase (*cya* gene), and therefore, unable to make 3′:5′ cyclic AMP (*c* AMP) or defective in the *c* AMP receptor protein (*crp* gene) are lysogenized at reduced frequency by phage P22. One class of the bacterial mutants with an altered RNA polymerase (*rif* gene) also is lysogenized with reduced frequency. *P22c*$^+$ forms clear plaques on all three mutant strains, suggesting that reduction is defective. Addition of *c* AMP to the adenyl cyclase mutants restores the ability of the phage to lysogenize these cells as evidenced by the formation of turbid plaques. The effect of the bacterial mutations appears to be on establishment rather than maintenance of lysogeny. The rare lysogens that are produced from the three classes of bacterial mutants are as stable as the lysogens of wild-type cells. In addition, a temperature-sensitive adenyl cyclase mutant can be lysogenized at 25° and it remains stable at 42°. HONG et al. (1971) propose that in wild-type cells the *c* AMP receptor protein in combination with *c* AMP activates the bacterial RNA polymerase to transcribe phage genes required for efficient lysogenization. Under conditions of strong catabolite repression, when the supply of energy and biosynthetic components is high and the level of *c* AMP is low, the phage would multiply and lyse the cell. The phage would lysogenize the cell when the supply of energy is deficient and the level of *c* AMP is high.

HONG et al. (1971) have also isolated phage P22 mutants which produce turbid plaques on the bacterial mutant strains and recover the ability to lysogenize them. These *cly*$^-$ mutants (for *c*ontrol of *ly*sogenization) lysogenize the wild-type bacteria at such high frequencies that they do not form plaques on this host and are defective mutants in the sense that none of them enter the lytic cycle. All *cly* mutants fall into one complementation group and all are recessive to *P22cly*$^+$. The *cly* gene maps in the *C* region in the order c_3-c_2-c_1-*cly-12*. The *cly*$^+$ function is required for the lytic response on infection of wild-type cells, but cells lysogenic for *cly*$^-$ prophage are induced by ultraviolet light to produce normal yields of phage.

These authors have also observed that phage P22 forms clear plaques on pyridoxine-requiring mutants of *S. typhimurium*. Addition of small amounts of pyridoxine to nutrient agar plates changes the plaque morphology to turbid.

B. Integration

CAMPBELL (1962) proposed that integration of prophage is achieved by a recombinational event between circularized phage DNA and the host chromosome. Induction is initiated by the reverse recombination, excising the prophage in a circular form. Closed circular duplex molecules of input and of newly synthesized phage P22 DNA have been observed in c^+ infections leading to lysogeny (RHOADES and THOMAS, 1968). Some 30 to 40 percent of the linear parental phage DNA is converted to circular forms from the 9th to 15th minute of the infection, after the repression of DNA synthesis is initiated. BOTSTEIN (personal communication) has detected closed circles of P22 DNA in intermediate I extracted from c^+ infected cells. Circularization may be achieved by genetic recombination between the terminally repetitious ends of the DNA molecule.

BOTSTEIN and MATZ (1970) report that *P22erf*$^-$ not only does not grow in *rec*$^-$ cells, but does not lysogenize them. However, *P22erf*$^-$ can become stable prophage in mixed infection of *rec*$^-$ cells with *erf*$^+$ phage. Thus, the *erf*$^+$ function is required for integration, but not to maintain lysogeny. Of particular interest is the finding that, once established as prophage, an *erf*$^-$ mutant can grow normally in a *rec*$^-$ host when induced. Since a circular phage genome is presumed to be the product of the prophage excision mechanism on induction, BOTSTEIN and MATZ interpret these findings to mean that circularization of the phage genome is essential to successful replication. In the absence of both the phage *erf* function and the host *rec* function, circularization cannot be accomplished on infection, resulting in both the inability to grow and to lysogenize.

SMITH and LEVINE (1967) isolated *int*$^-$ mutants of phage P22 (originally called *L* mutants) which cannot carry out a late step necessary for prophage integration. This type of mutation does not interfere with phage replication. Under conditions which lead to high frequencies of lysogenization with wild-type, *P22int*$^-$ infections give normal levels of host survival but not as lysogens. Reduction is effective, but the phage genomes are progressively diluted out among the progeny cells. *P22int*$^-$ can become prophage by complementation and these lysogens are stable. The *int* gene product is then necessary for integration but not for maintenance of the lysogenic state. Experiments with temperature-sensitive *int*$^-$ mutants show that the gene product acts late, after the infected cells begin to divide, and does not complete its activity in all cells until at least the third generation postinfection. By analogy with the *int*$^-$ mutants of phage λ (WEIL and SIGNER, 1968; ECHOLS et al., 1968; SIGNER et al., 1969; ECHOLS, 1970), the *int* gene is thought to be the structural gene for a specific integrase governing recombination between the circularized phage genome and the host chromosome.

Common functions appear to be required for phage DNA synthesis and for prophage integration (LEVINE and SCHOTT, 1971). As already mentioned, at high temperature, *P22ts18* and *P22ts12* fail to synthesize phage DNA in single infection, but complement for DNA synthesis and phage production in mixed infection with each other and with non-mutant phage. Infections with *P22ts12c*$^+$ or *P22ts18c*$^+$ at high temperature give cell survival as sensitives; lysogeny is not established. These mutants behave like *int*$^-$ mutants. Paradoxically, *P22ts12c*$^+$ and *P22ts18c*$^+$ do not complement for lysogenization in mixed infection, and furthermore, each of these mutants interferes with lysogenization by $P22c^+$. These mutants behave as dominant to wild-type for lysogenization. The inability of either mutant phage to lysogenize at 37° is not due to failure to synthesize the early peak of phage DNA observed in wild-type infections, since mixed infections with both mutants or either mutant and wild-type, which complement for DNA synthesis, also fail to give rise to lysogens. This implies that functional gene *18* and *12* products are required in a reaction (or reactions) for the establisment of lysogeny distinct from that in replication of the phage genome. The mutational lesions clearly affect events leading to integration of the prophage rather than maintenance of the lysogenic state. Lysogens carrying either $P22ts12c^+$ or $P22ts18c^+$ prophage are readily formed at 25°. On shifting to 37°, these are as stable as wild-type lysogens.

The pleiotropic action of gene *18* and *12* products might be understood in the following way. Suppose genes *18* and *12* coded, respectively, for proteins with polymerase and ligase activities. Mutational lesions would be expected to interfere with phage replication. The same enzymatic activities might be required to accomplish the covalent bonding needed to make closed circular phage DNA and/or to integrate the prophage. The same mutational lesions could then prevent establishment of prophage at one or the other step. The negative complementation for lysogeny raises the possibility that we are dialing with one locus and that the synthesis of phage DNA in mixed infection may be the consequence of intragenic complementation. Despite this reservation, we will assume that the *ts 12* and *ts 18* mutations represent two different genes. Negative complementation suggests that each gene product is multimeric.

Genes *18* and *12* are closely linked, mapping to the right of the *C* region, and appear to be analogous to genes *O* and *P* of phage λ in map position and in fuction in phage DNA replication. However, mutations in genes *O* and *P* do not interfere with lysogenization by phage λ (BROOKS, 1963; BROWN and ARBER, 1964; SIGNER, 1970; FREIFELDER, personal communication). Two "early" essential genes, cistrons *A* and *B*, have been identified in P2 by LINDAHL (1969, 1970). Infection with phage mutant in either of these genes under nonpermissive conditions does not result in killing of the bacterial host, which may instead become lysogenic.

ZINDER (1958) described a mutant, V_1, which is able to initiate reduction. Infected cells begin to divide, segregation of sensitive cells takes place but stable integration fails. Only after a number of divisions do cells still carrying phage genomes produce phage progeny. GOUGH (1968) re-examined the V_1 and similar mutations and concluded that this locus was needed to maintain the prophage and not for the process of integration itself. He renamed this locus *mnt* for *maint*enance of lysogeny (see section Vc).

Infections under nonpermissive conditions with phage mutants known to affect head formation, tail part production, or cell lysis show normal frequencies of lysogeny.

Integration of the P22 prophage is sensitive to temperature (SMITH and LEVINE, 1967). The proportion of cells carrying the wild-type prophage among the progeny of surviving infected cells can vary from as much as 90 percent for infections carried out at 25° to less than 1 percent at 42°. The ratio of phage carriers to sensitive cells for infections at 37° is usually around 1.

The site of integration of P22 prophage, between the *proA* and *proC* genes of the host, is highly specific (ZINDER, 1963; SMITH and STOCKER, 1966; SMITH and LEVINE, 1965; YOUNG and HARTMAN, 1966; SMITH, 1968). Cells deleted in this region can be lysogenized by P22, but at a frequency orders of magnitude lower than for cells which have the normal attachment region (RAO and SMITH, 1968). Normal integration can be visualized as a recombinational event between the host chromosome attachment region and an attachment region on the circular phage DNA lying between the *int* and gene *9* loci giving the structure shown in Fig. 2. The genetic control of attachment specificity is not clear, but probably resides in the base sequences at the attachment sites. The *C* region plays no determinative role in prophage location. Phage P221 is a P22-related phage released from *Salmonella typhimurium* strain-LT2 infected with P22 (YAMAMOTO and ANDERSON, 1961; YAMAMOTO and WEIR, 1966a and b; YOUNG et al., 1966). Phage P221 shares a common *C* region with P22 (YAMAMOTO and WEIR, 1966b), yet P221 prophage is integrated between a cluster of trytophan and a cluster of galactose genes on the host chromosome (YOUNG and HARTMAN, 1966).

C. Maintenance and Induction

P22 lysogens are, as a rule, exceedingly stable. On the average, one cell in 1 000 per division cycle is spontaneously induced to produce progeny phage. P22 lysogens are readily induced with high frequency by exposure to such agents as ultraviolet light, mitomycin C and streptonigrin (LEVINE, 1961; LEVINE and BORTHWICK, 1963). Phage production and lysis in practically every cell follows high temperature exposure of lysogens carrying prophages with either *ts*c_2 (LEVINE and SMITH, 1964) or *tsmnt* (GOUGH, 1968) mutant alleles. The structural genes for the *mnt* and c_2 repressors are distantly linked (ZINDER, 1958; GOUGH, 1968) (Fig. 1). The continuous action of both gene products is required for maintenance of prophage. The kinetics of thermal induction of *tsmnt* (GOUGH, 1968) and *ts*c_2 prophages are very different. Colony-forming cells decrease in number and infective centers appear from time-zero of a shift of cells carrying the *ts*c_2 prophage from 30 to 43°. By 20 minutes at high temperature, induction is complete; all cells are converted to infective centers. After shifting cells lysogenic for the *tsmnt* prophage from 30 to 40°, the cell number increases three-fold and infective centers are not completely induced until 90–120 minutes after the shift. Long filamentous bacterial forms are found at high temperature in which phage DNA is synthesized between 90 and 120 minutes (GOUGH, unpublished results). Lysogens carrying the double mutant prophage, *ts*c_2*tsmnt*, are induced at 43° with the kinetics of *ts*c_2 lysogens (GOUGH, 1968). *Mnt*$^+$ is dominant to *mnt*$^-$. The data are consistent with the hypothesis that removal of *mnt* product results in the eventual release from c_2 repression.

Successful phage production after induction requires not only inactivation of the c_2 or *mnt* repressors, but the normal activities of at least two other loci, the *int* and *xis* loci. As already mentioned, *P22int*$^-$ can be complemented with *P22int*$^+$ to give *int*$^-$ lysogens. Once established, the *int*$^-$ prophage is stable in the absence of *int*$^+$ function. Ultraviolet induction of *int*$^-$ lysogens or thermal induction of *ts*c_2*int*$^-$ lysogens leads to phage production and cell lysis (SMITH, 1968). However, the vast majority of the particles are defective. The defective particles contain a normal amount of DNA but the DNA is mainly bacterial in origin and appears to be derived from the *proC* region adjacent to one end of the prophage. The defective lysates transduce the *proC* marker with high frequency relative to other bacterial markers. The representation of phage markers in the defective particles is strongly polar, with markers close to the *proA* end of the bacterial chromosome being rarely present. It appears that wild-type *int* function is required not only for efficient integration of the prophage, but for normal excision of the prophage genome after induction.

SMITH (personal communication) has isolated mutations in a second gene, *xis*. The product of this locus must be expressed along with that of the *int* locus for excision of prophage P22.

In contrast to the results described above for thermal induction of a *tsmnt* lysogen, high temperature treatment of a *tsmntint*$^-$ lysogen does not result in death of the cell and does not result in phage production. The cell survives as a stable lysogen. That is, normal *int* function is needed for *mnt* induction. GOUGH (1970) suggests that the *mnt* locus regulates the *int* locus.

RAY and BURMA (1970) reported that penicillin-induced spheroplasts absorb P22 normally and show single-step growth curves similar to those of infected whole cells except that the burst sizes are lower. However, induction of phage production could not be demonstrated in spheroplasts of cells lysogenic for a *ts*c_2 prophage. Surprisingly, spheroplasts of lysogens lose immunity to superinfection by homologous nonvirulent phage (CHAKRAVORTY and LEVINE, 1970). The ability to support phage growth suggests that the inability to induce phage in these spheroplasts is due to an earlier step, perhaps excision of the prophage.

Wild-type phage P22 can integrate in *rec*$^-$ bacteria and heat induction of *rec*$^-$ cells lysogenic for *ts*c_2 prophage is normal (WING, 1968). The bacterial Rec function, then, is not directly involved in the lysogenic phenomenon. *Rec*$^-$ lysogens cannot be induced by ultraviolet irradiation or treatment with streptonigrin or mitomycin C (WING, 1968), presumably because the cell lacks part of the mechanism required to inactivate repressor.

VI. Superinfection Immunity

The immunity of a lysogen to superinfection is specific. Phages P22 and L are closely related (BEZDEK and AMATI, 1967; FARVE et al., 1968). They have similar morphologies, similar serological properties and show sufficient genetic homology to permit some recombination between them. Yet they differ in the immunity properties they impose on lysogens. Phage P22 is productive in an L lysogen, but phage L is repressed. Conversely, P22 lysogens will support the growth of phage L but not P22. Phages L and P22 are said to be *hetero-immune*. Control of immunity specificity is bipartite, involving the *C* region and the *mnt* region (Fig. 1) (BEZDEK and AMATI, 1968). Genetic crosses between phages P22 and L yield recombinant types which are still temperate but which have altered immunity specificities. Analysis of these recombinants indicates changes in immunity specificity arise if the *C* region of one parent and the *I* (for *i*mmunity) region of the second form a new combination. The *immI* region is closely linked to the plaque morphology marker m_3 and therefore identical or very closely linked to *mnt*. These two regions affect both the immunity produced in a lysogen and the sensitivity to it of the superinfecting phage. The bipartite control in phage P22 is different from the uniregional control of immunity specificity by the *C* region in the λ system (KAISER and JACOB, 1957).

VII. Virulent Mutants

Mutants of phage P22 have been isolated which form plaques on P22 lysogens (BRONSON and LEVINE, 1971a and b; SWANSON and BOTSTEIN, unpublished). These virulent phages can be classified into three groups. 1. *VirA* mutants arise spontaneously and form a single locus which maps near the *mnt* locus in the *immI* region. *VirA* mutants induce prophage with high frequency suggesting they grow in lysogens by virtue of inactivation of the repressor mechanism (SWANSON and BOTSTEIN, unpublished). 2. *VirB* mutants do not arise spontaneously and have been isolated only from mutagenized phage stocks. These are comprised of two distinct mutations in the *immC* region (BRONSON and LEVINE, 1971a): *K5*, which is a site specific mutation in the c_2 repressor gene, and *Vx*, a locus which maps in the region between the c_2 and c_3 genes (Fig. 1). 3. *VirC* mutants are triple mutants made up of the determinants of both VirA and VirB virulents. It is significant that these mutations are closely linked to the loci involved in immunity control and repressor synthesis.

One of the VirB mutants, *P22 vir B-3*, has been analyzed in detail (BRONSON and LEVINE, 1971a and b). At low multiplicities of infection, *P22 vir B-3* grows equally well in either a P22 lysogen or a sensitive host. Productive superinfection with *P22 vir B-3* does not cause induction of the prophage, suggesting that virulence is due to insensitivity to repression rather than inactivation of repressor. *P22vir B-3* is, however, not totally insensitive to repression. The burst size of the virulent mutant is greatly decreased in mixed infection of nonlysogenic cells with phage that produce c_2 repressor. In addition, the *K5* mutation which maps near the right end of the c_2 locus does not completely interfere with the production of active c_2 repressor, since the yield of progeny phages is greatly reduced as the multiplicity of infection of *P22vir B-3* increases in either sensitive or lysogenic host cells. This *multiplicity repression* disappears if an additional c_2 mutation is introduced into the virulent genome. The *mnt* repressor is not involved in multiplicity repression. We conclude that *vir B-3* synthesizes active c_2 repressor and exhibits a reduced but finite sensitivity to the action of the repressor.

Phages carrying either the *K5* or the *Vx* mutation are not in themselves virulent and do not show extensive replication of DNA in single infection of a lysogen. However, they complement each other in mixed superinfection of a lysogen, resulting in phage production. The data are consistent with *K5* and/or *Vx* being operator constitutive mutations under c_2 repressor control.

Preliminary data (BRONSON and LEVINE, unpublished) suggest that *P22K5* can constitutively express the functions of genes *18* and *12* in a lysogen. Mixed superinfection of lysogenic cells at nonpermissive temperature with *P22K5* and either *P22vir B-3 ts18* or *P22vir B-3ts12* give significantly increased levels of phage production over the appropriate controls. In contrast, *P22Vx* does not comple- ment for these mutant alleles and in addition does not appear to express the functions of genes *21, 6, 25* and *9* in a lysogen. Since a *Vx* function is needed for DNA synthesis in mixed infection in a lysogen with *P22K5* (BRONSON and LEVINE, 1971b), it is probable that *P22Vx* is constitutive for an as yet unidentified gene or genes necessary for replication of the phage genome. Further, stable lysogens for Vxc^+ have not yet been observed (BRONSON and LEVINE, unpublished). $P22Vxc^+$ can, however, complement $P22c_1$ to be integrated as prophage. These results suggests that the *Vx* mutation causes a deficiency in maintenance of lysogeny, which may be the result of constitutive expression of gene functions that are normally under repressor control.

VIII. Replication Inhibition

Phage P22 is sensitive to replication inhibition (LEVINE et al., 1970), first described for phage λ and P2 by THOMAS and BERTANI (1964). In mixed superinfection of a lysogen by *P22vir B-3* and the nonvirulent $P22c_1$, the yield of the nonvirulent phage is less than its input, while virulent progeny represent greater than *98* percent of the yield. The *vir B-3* DNA associates with the phage replication complex and undergoes normal replication and maturation. The nonvirulent DNA does not associate with the complex in the mixed superinfection and $P22c_1$ remains repressed even though the virulent replicates normally. Only after inactivation of repressor do both types of DNA bind to intermediate I and both phages are produced in equal amounts. Association with the replication machinery in the lysogen is thus under immunity control.

The inability of the virulent mutant to complement nonvirulent DNA into the active replication machinery suggests that repressor plays a direct physical role in replication inhibition. Repressor may exert its *cis* effect by blocking the site for binding to the replication machinery. Alternatively, repressor may block transcription of a critical region of each DNA molecule which is a prerequisite for binding to the machinery. DOVE et al. (1969), have suggested the need for transcription of a particular region of the λ genome in release from replication inhibition. It should be remembered that gene *25* product is needed to demonstrate binding of input DNA to intermediate I.

Since escape from replication inhibition is a prerequisite for phage development in the presence of repressor, it follows that at least one component mutation should confer this *cis* property on *P22virB-3*. P*22K5* completely escapes from replication inhibition. In mixed superinfections of P22 lysogens with *virB-3, K5* progeny are at least as common as virulent particles (BRONSON and LEVINE, 1971b). In accord with expectations, *K5* genomes associate with intermediate I in a lysogen (BRONSON, unpublished data). *P22Vx* gives only partial escape from replication inhibition (BRONSON and LEVINE, 1971b).

IX. Conversion and Superinfection Exclusion

A number of P22 loci are known, other than the *mnt* and c_2 repressor loci, whose activities are expressed in the prophage state.

P22 has the ability to elicit somatic antigen 1 in the host cell. The antigen is produced both when the phage is in the vegetative state and in the prophage state (STOCKER, 1958; ZINDER, 1958; YOUNG, et al., 1964). P22 carrying the a_1 mutation cannot induce this conversion (YOUNG et al., 1964). The exact location of the a_1 mutation is not known but it maps between m_3 and c_2 on the phage linkage map.

Wild-type P22 lysogens prevent growth of superinfecting P22 not only by the immunity system, but also by an exclusion mechanism (WALSH and MEYNELL, 1967; RAO, 1968). Superinfecting P22 cannot replicate in the lysogen, cannot complement nor recombine with the prophage genome even if immunity is released by induction (RAO, 1968). Related heteroimmune phages L, MG178 and Mg40 are also excluded in P22 lysogens. The exclusion system also acts against host DNA as it is responsible for inhibition of transduction in P22 lysogens (EBEL-TSIPIS and BOTSTEIN, 1971).

Mutants of P22, called *sie*$^-$ mutants for *s*uper*i*nfection *e*xclusion have been isolated (WALSH and MEYNELL, 1967; RAO, 1968; SUSSKIND et al., 1971). *Sie*$^-$ lysogens plate heteroimmune phage and give superinfecting P22 progeny particles on induction. The exclusion is controlled by two P22 loci, *sieA* and *sieB*, located in the P22 linkage map near the *mnt* locus and *C* regions, respectively (SUSSKIND et al., 1971). The A exclusion system excludes phages P22, MG178, L and transducing particles. The B system acts only on the heteroimmune phages. The two exclusion systems act independently of each other and of the immunity specificity of the prophage.

The mechanism of action of the A exclusion system is not clear. Differences between *sieA*$^-$ and *sieA*$^+$ lysogens cannot be explained by differences in phage adsorption, injection and degradation of infected DNA (RAO, 1968; SUSSKIND et al., 1971). However, the route of entry of the DNA appears critical. Only DNA which is injected into an excluding lysogen from a phage particle is excluded. Phage DNA or bacterial DNA which enters a $sieA^+$ lysogen by conjugation is not excluded (RAO, 1968). Further, DNA from *P22vir B-3* associates with the phage replication complex in a sie^- lysogen, but not in a sie^+ lysogen (DARLINGTON and LEVINE, 1971).

The linkage relationships in the prophage (Fig. 2) suggest that these genes may be part of two constitutive operons; one containing *mnt*, *sieA*, and a_1 and the other c_2 and *sieB*.

X. Effects on Host Cell Metabolism

Profound changes in host cell macromolecular synthesis have been described for a variety of phage infections. Survival as a lysogen would require that any effect a temperate phage may have on its host must be potentially reversible. Reversible repression of host cell protein and DNA synthesis has been described for P22. Almost instantly on infection, host DNA synthesis is inhibited and in the case of $P22c^+$ infections the inhibition is reversed at a very precise time: 15–16 minutes after infection (SMITH and LEVINE, 1965a; LEVINE and SCHOTT, 1971).

Electrophoresis analysis of proteins synthesized during $P22c^+$ infection show decreased synthesis of certain host-specific proteins early in infection followed by subsequent recovery of synthesis. The release of inhibition is not observed in infections leading to phage production (COHEN and LEVINE, 1966). The amount of host cell nuclease activity specific for denatured DNA decreases after infection (ISRAEL et al., 1971; in press). This decrease reaches its maximum 10 minutes after infection and varies from 50 to 25 percent of the original level. In $P22c^+$ infections, the nuclease activity recovers rapidly after the initial drop, but the recovery is not observed in productive $P22c_1$ infections. The control of these alterations clearly rests with the phage, since these effects are not found when a lysogenic cell is used as host or when $P22ts25c_1$ is the infecting phage at high temperature. No specific phage genes have yet been implicated. The induction of host cell L-arabinose isomerase is repressed on infection with $P22c_1$ or $P22c_2$. This repression is reversible in infections leading to lysogeny (CHAKRAVORTY, 1970).

XI. Closing Comments

The bipartite control of lysogeny is a unique feature of phage P22. Both the *immI* and *immC* regions of the phage genome are involved as shown by the following observations. 1. Maintenance of the prophage requires the activities of both the c_2 gene in the *immC* region and the *mnt* locus in the *immI* region. 2. Virulence can be achieved by mutations in either the *immI* region or the *immC* region. 3. Recombinants between heteroimmune phages have altered immunity specificity if they carry the *immC* region of one parent and the *immI* region of the second.

We suggest that the dual control can be understood by the following considerations. The *immC* region is responsible for both establishment and maintenance of lysogeny. All three *C* genes are needed for reduction but the continuous functioning of the c_2 gene is needed for maintenance. The c_2 repressor plays the primary role in repressing phage replication and transcription through two operator sites located to either side of the c_2 locus. The *immI* region codes for the synthesis and control of a gene product which either destroys the c_2 repressor or inhibits its synthesis. The synthesis of this *anti-repressor*, from an as yet unidentified structural gene, is regulated by the *mnt* gene product, a repressor of an anti-repressor acting through a closely linked operator site. The role of the *immI* region is then an indirect one by virtue of its ability to antagonize c_2 repression, rather than by direct action on essential gene functions. This formulation allows plausible interpretations of various experimental results. Only a few of these will be discussed here as examples.

1. VirA mutants induce prophages in superinfected lysogens. VirA phages may be mutant in either the promoter or operator sites for the anti-repressor. The resultant constitutive synthesis of anti-repressor and subsequent release from c_2 repressor control permits both the virulent and prophage genomes to replicate. 2. Phage production is delayed for a few divisions when a *tsmnt* lysogen is induced at high temperature. It may take that long to synthesize anti-repressor after inactivation of the *mnt* repressor. ZINDER (1958) reported that on infection of sensitive cells with *mnt*⁻ phage, the cells survive and begin to divide and only after a number of divisions are phage produced. Anti-repressor may not be made early in infection. 3. The altered immunity specificity of recombinants between P22 and L can be explained by making some additional assumptions about the interactions of the specific *immI* and *immC* regions.

A regulator of repressor synthesis has been described in the λ system (EISEN et al., 1970; SPIEGELMAN et al., 1970). The repressor regulator is capable in *trans* of preventing expression of gene c_1, the structural gene for the λ repressor. Mutations removing this function map in the *cro* gene located in the λ immunity region.

The above outlined hypothesis to explain the bipartite control of lysogeny by phage P22 suggests a number of experimental tests which remain to be carried out.

Acknowledgments. This paper ist dedicated to Professor TRACY M. SONNEBORN of the Department of Zoology at Indiana University on the occasion of the 35th anniversary of his discovery of mating types in *Paramecium aurelia* and his 67th birthday. The role he played in my scientific training and the warmth of friendship he and Mrs. SONNEBORN extended to all his students are fondly remembered.

I would like to thank the whole fraternity of P22 workers for the open and collabora tive environment in which we work. My special thanks to Drs. VANCEISRAEL, DAVID BOTSTEIN, MARY WOODWORTH-GUTAI, GAIL WERTZ, MORLEY BRONSON, DAVID FRIEDMAN, Miss BARBARA HOFFMAN and Mr. STEVE WEAVER for many fruitful and heated discussions during the preparation of this review.

The work of my laboratory has been supported by U. S. Public Health Service grants GM-1519-04 and NIH-5-TO1-GM00071 from the National Institute of General Medical Sciences.

References

Bezdek, M., Amati, P.: Properties of P22 and a related *Salmonella typhimurium* phage. I. General features and host specificity. Virology **31**, 272–278 (1967).

Bezdek, M., Amati, P.: Evidence for two immunity regulator systems in temperate bacteriophages P22 and L. Virology **36**, 701–703 (1968).

Bezdek, M., Soska, J.: Evidence for an early regulatory function in phage P22. Molec. gen. Genet. **108**, 243–248 (1970).

Botstein, D.: Synthesis and maturation of phage P22 DNA. I. Identification of intermediates. J. molec. Biol. **34**, 621–641 (1968).

Botstein, D., Levine, M.: Synthesis and maturation of phage P22 DNA. II. Properties of temperature-sensitive phage mutants defective in DNA metabolism. J. molec. Biol. **34**, 643–654 (1968a).

Botstein, D., Levine, M.: Intermediates in the synthesis of phage P22 DNA. Cold Spr. Harb. Symp. quant. Biol. **33**, 659–667 (1968b).

Botstein, D., Matz, M. J.: A recombination function essential to the growth of bacteriophage P22. J. molec. Biol. **54**, 417–440 (1970).

Boyd, J. S. K.: Observations on the relationship of symbiotic and lytic bacteriophage. J. Path. Bact. **63**, 445–457 (1951).

Bronson, M. J., Levine, M.: Virulent mutants of phage P22. I. Isolation and genetic analysis. J. Virol. **7**, 559–568 (1971a).

Bronson, M. J., Levine, M.: Virulent mutants of phage P22. II. Physiological analysis of P22 *virB-3* and its component mutations. Virology, in press (1971b).

Brooks, K.: Studies in the physiological genetics of some suppressor-sensitive mutants of bacteriophage λ. Virology **26**, 489–499.

Brown, A., Arber, W.: Temperature-sensitive mutants of coliphage lambda. Virology **24**, 237–239 (1964).

Campbell, A.: Episomes. Advanc. Genet. **12**, 101–145 (1962).

Chakravorty, M.: Induction and repression of L-arabinose isomerase in bacteriophage infected *Salmonella typhimurium*. J. Virol. **5**, 541–547 (1970).

Chakravorty, M., Levine, M.: Absence of immunity to superinfection in spheroplasts of a lysogen. J. Virol. **5**, 264–265 (1970).

Cohen, L. W.: Delayed lysis with a mutant of *Salmonella* bacteriophage P22. J. Virol. **4**, 209–213 (1969a).

Cohen, L. W.: Delayed lysis with *Salmonella* bacteriophage P22: Induction of lysis by addition of cysteine or histidine to the growth medium. J. Virol. **4**, 214–218 (1969b).

Cohen, L. W., Levine, M.: Detection of proteins synthesized during the establishment of lysogeny with phage P22. Virology **28**, 208–213 (1966).

Darlington, O. F., Levine, M.: Superinfection exclusion by P22 prophage and the replication complex. J. Virol. **8**, 347–348 (1971).

Dove, W. F., Hargrove, E., Ohashi, M., Guha, A.: Replicator activation in lambda. Jap. J. Genet. **44**, 11–22 (1969).

Ebel-Tsipis, J., Botstein, D.: Superinfection exclusion by P22 prophage in lysogens of *Salmonella typhimurium*. I. Exclusion of generalized transducing particles. Virology **45**, 629–637 (1971).

Echols, H.: Integrative and excisive recombination by bacteriophage λ: Evidence for an excision-specific recombination protein. J. molec. Biol. **47**, 575–583 (1970).

Echols, H., Gingery, R., Moore, L.: Integrative recombination function of bacteriophage λ: Evidence for a site-specific recombination enzyme. J. molec. Biol. **34**, 251 (1968).

Eisen, H., Brachet, P., Pereira da Silva, L., Jacobs, F.: Regulation of repressor expression in λ. Proc. nat. Acad. Sci. (Wash.) **66**, 855–862 (1970).

Favre, R., Amati, P., Bezdek, M.: Properties of P22 and a related *Salmonella typhimurium* phage. II. Effect of H markers on infection of strain 1559. Virology **35**, 238–247 (1968).

Gilbert, W., Dressler, D.: DNA replication: The rolling circle model. Cold Spr. Harb. Symp. quant. Biol. **33**, 473–484 (1968).

Gough, M.: Second locus of bacteriophage P22 necessary for the maintenance of lysogeny. J. Virol. **2**, 992–998 (1968).

Gough, M.: Requirement for a functional *int* product in temperature induction of prophage P22 *tsmnt*. J. Virol. **6**, 320–325 (1970).

Gough, M., Levine, M.: Circularity of the phage P22 linkage map. Genetics **58**, 161–169 (1968).
Hong, J. S., Smith, G. R., Ames, B. N.: The cyclic-AMP (3′:5′) level of the bacterial host regulates the viral decision between lysogeny and lysis. Proc. nat. Acad. Sci. (Wash.) **68**, 2258–2262 (1971).
Israel, V.: The production of inactive phage P22 particles following induction. Virology **33**, 317–322 (1967).
Israel, V., Anderson, T. F., Levine, M.: In vitro morphogenesis of phage P22 from heads and base-plate parts. Proc. nat. Acad. Sci. (Wash.) **57**, 284–291 (1967).
Israel, V., Woodworth-Gutai, M., Levine, M.: Phage induced inhibition of host exonuclease activity. Bact. Proc., 197 (1971).
Kaiser, A. D., Jacobs, F.: Recombination between related temperate bacteriophages and the genetic control of immunity and prophage localization. Virology **4**, 509–521 (1957).
Kolstad, R. A., Prell, H. H.: An amber map of *Salmonella* phage P22. Molec. gen. Genet. **104**, 339–350 (1969).
Levine, M.: Mutations in the temperate phage P22 and lysogeny in *Salmonella*. Virology **3**, 22–41 (1957).
Levine, M.: Effect of mitomycin C in interactions between temperate phages and bacteria. Virology **13**, 493–499 (1961).
Levine, M., Borthwick, M.: The action of streptonigrin on bacterial DNA metabolism and on induction of phage production in lysogenic bacteria. Virology **21**, 568–574 (1963).
Levine, M., Chakravorty, M., Bronson, M. J.: Control of the replication complex of bacteriophage P22. J. Virol. **6**, 400–405 (1970).
Levine, M., Curtiss, R.: Genetic fine structure of the C region and the linkage map of phage P22. Genetics **46**, 1573–1580 (1961).
Levine, M., Schott, C.: Mutations of phage P22 affecting bacteriophage DNA synthesis and lysogenization. J. molec. Biol., **62**, 53–64 (1971).
Levine, M., Smith, H. O.: Sequential gene action in the establishment of lysogeny. Science **146**, 1581–1582 (1964).
Lew, K. K., Roth, J. R.: Isolation of UGA and UAG nonsense mutants of bacteriophage P22. Virology **40**, 1059–1062 (1970).
Lindahl, G.: Genetic map of bacteriophage P2. Virology **39**, 839–860 (1969).
Lindahl, G.: Bacteriophage P2: Replication of the chromosome requires a protein which acts only on the genome that coded for it. Virology **42**, 522–533 (1970).
Lwoff, A.: Lysogeny. Bact. Rev. **17**, 269–337 (1953).
Rabovsky, D., Konrad, M.: Gene expression in bacteriophage λ. I. The kinetics of the requirement for N gene product. Virology **40**, 10–17 (1970).
Rao, R. N.: Bacteriophage P22 controlled exclusion in *Salmonella typhimurium*. J. molec. Biol. **35**, 607–622 (1968).
Rao, R. N., Smith, H. O.: Phage P22 lysogens of a *Salmonella typhimurium* mutant deleted at the normal prophage attachment site. Virology **36**, 328–330 (1968).
Ray, R. K., Burma, D. P.: Multiplication of phage P22 in penicillin spheroplasts of *Salmonella typhimurium*. J. Virol. **5**, 45–50 (1970).
Rhoades, M., MacHattie, L. A., Thomas, C. A., Jr.: The P22 bacteriophage DNA molecule. I. The mature form. J. molec. Biol. **37**, 21–40 (1968).
Rhoades, M., Thomas, C. A., Jr.: The P22 bacteriophage DNA molecule. II. Circular intracellular forms. J. molec. Biol. **37**, 41–61 (1968).
Signer, E. R.: On the control of lysogeny in phage λ. Virology **40**, 624–633 (1970).
Signer, E. R., Weil, J., Kimball, P. C.: Recombination in bacteriophage λ. III. Studies on the nature of the prophage attachment region. J. molec. Biol. **46**, 543–563 (1969).
Smith, H. O.: Defective phage formation by lysogens of integration deficient phage P22 mutants. Virology **34**, 203–223 (1968).
Smith, H. O., Levine, M.: Two sequential repressions of DNA synthesis in the establishment of lysogeny by phage P22 and its mutants. Proc. nat. Acad. Sci. (Wash.) **52**, 356–363 (1964).
Smith, H. O., Levine, M.: The synthesis of phage and host DNA in the establishment of lysogeny. Virology **25**, 585–590 (1965a).
Smith, H. O., Levine, M.: Gene order in prophage P22. Virology **27**, 229–231 (1965b).

SMITH, H. O., LEVINE, M.: A phage P22 gene controlling integration of prophage. Virology **31**, 207–216 (1967).
SMITH, S. M., STOCKER, B. A. D.: Mapping of prophage PLT22 in *Salmonella typhimurium*. Virology **28**, 413–419 (1966).
SPIEGELMAN, W. G., HEINEMANN, S. F., BRACHET, P., PEREIRA DA SILVA, L., EISEN, H.: Regulation of the synthesis of phage lambda repressor. Cold Spr. Harb. Symp. quant. Biol. **35**, 325–330 (1970).
STOCKER, B. A. D.: J. gen. Microbiol. **18**, ix (1958a).
STOCKER, B. A. D.: Abstr. Intern. Congr. Microbiol. 7th Congr., 65 (1958b).
STREISINGER, G., EDGAR, R. S., DENHARDT, G. H.: Chromosome structure in phage T4. I. Circularity of the linkage map. Proc. nat. Acad. Sci. (Wash.) **51**, 775–779 (1964).
SUSSKIND, M. M., WRIGHT, A., BOTSTEIN, D.: Superinfection exclusion by P22 prophage in lysogens of *Salmonella typhimurium*. II. Genetic evidence for two exclusion systems. Virology **45**, 638–652 (1971).
THOMAS, R., BERTANI, L. E.: On the control of the replication of temperate bacteriophages superinfecting immune hosts. Virology **24**, 241–253 (1964).
WALSH, J., MEYNELL, G. G.: The isolation of nonexcluding mutants of phage P22. J. gen. Virol. **1**, 581–582 (1967).
WEIL, J., SIGNER, E. R.: Recombination in bacteriophage λ. II. Site-specific recombination in bacteriophage λ. J. molec. Biol. **34**, 273–279 (1968).
WING, J. P.: Integration and induction of phage P22 in a recombination-deficient mutant of *Salmonella typhimurium*. J. Virol. **2**, 702–709 (1968).
WING, J. P., LEVINE, M., SMITH, H. O.: Recombination-deficient mutant of *Salmonella typhimurium*. J. Bact. **95**, 1828–1834 (1968).
WOODWORTH-GUTAI, M., ISRAEL, V., LEVINE, M.: A new exonuclease activity after phage P22 infection. Bact. Proc., 210 (1971).
YAMAGAMI, H., YAMAMOTO, N.: Contribution of the bacterial recombination function to replication of bacteriophage P22. J. molec. Biol. **53**, 281–285 (1970).
YAMAMOTO, N., ANDERSON, T. F.: Genomic masking and recombination between serologically unrelated phage P22 and P221. Virology **14**, 430–439 (1961).
YAMAMOTO, N., WEIR, M. L.: Genetic relationships between serologically unrelated bacteriophages P22 and P221b. Virology **28**, 168–169 (1966a).
YAMAMOTO, N., WEIR, M. L.: Boundary structure between homologous and nonhomologous regions in serologically unrelated bacteriophages P22 and P221. I. Mapping by recombination found in the cultures of double lysogenic strains for P22*h* and P221. Virology **28**, 324–329 (1966b).
YOUNG, B. G., FUKAZAWA, Y., HARTMAN, P. E.: A P22 bacteriophage mutant defective in antigen conversion. Virology **23**, 279–283 (1964).
YOUNG, B. G., HARTMAN, P. E.: Sites of P22 and P221 prophage integration in *Salmonella typhimurium*. Virology **28**, 265–270 (1966).
YOUNG, B. G., HARTMAN, P. E., MOUNDRIANAKIS, E. N.: Some phages released from P22-infected *Salmonella*. Virology **23**, 249–264 (1966).
ZINDER, N. D.: Infective heredity in bacteria. Cold Spr. Harb. Symp. quant. Biol. **28**, 261–269 (1953).
ZINDER, N. D.: Lysogenization and superinfection immunity in *Salmonella*. Virology **5**, 291–326 (1958).
ZINDER, N. D.: Discussion. In: Methodology in basic genetics. San Francisco: Holden-Day 1963.
ZINDER, N. D., LEDERBERG, J.: Genetic exchange in *Salmonella*. J. Bact. **64**, 679–699 (1952).

The Morphogenesis of Poliovirus

BRUCE A. PHILLIPS[1]

With 4 Figures

Contents

I. Introduction

Poliovirus is one of the simplest and most extensively studied animal viruses. Yet, in spite of the voluminous literature, many basic and important questions concerning its morphogenesis remain unanswered. For example, what mechanisms insure that the progeny virus retain the precise structural features of the infecting particle? How does the viral RNA combine with the capsid proteins to form the virion and how are host RNA molecules excluded from this process? Do non-capsid viral proteins influence these events? Thus, while infectious tobacco mosaic virus particles have been reconstituted *in vitro* from viral RNA and protein monomers, both obtained by controlled degradation of the virion, it remains to be shown that this is in fact the mechanism of morphogenesis *in vivo*.

Because an understanding of the fine structure of the virion is crucial to an understanding of the mechanisms governing its morphogenesis, a description of the poliovirion has been included.

No attempt has been made to provide for a complete bibliography and much relevant material has been omitted on a highly subjective and selective basis. The purpose of this review is to describe the evolution of current ideas

1 I acknowledge the many helpful criticisms of Drs. J. S. YOUNGNER and H. THACORE during the preparation of this review. This review was aided by Award No. AI-08368 from the National Institutes of Health.

concerning poliovirus replication and to present a reasonably detailed picture of the morphogenesis of poliovirus consistent with past and recent experimental findings. The validity of proposed models will de determined by future experiments.

II. The Poliovirion

A. Morphology

Polioviruses, which are members of the picornavirus family, are small ribonucleoprotein particles having a diameter of about 280 A (McFERRAN et al., 1971). There is no membraneous material associated with the purified virus particle. The particle weight, based on an RNA content of 30%, probably is closer to 8×10^6 than the previously accepted value of 7×10^6 daltons (see below).

When negatively stained with phosphotungstic acid (PTA) and examined in the electron microscope, the great majority of poliovirus particles exhibit a distinct hexagonal profile as shown in Fig. 1 (also see HÖGLUND, 1968; McFERRAN et al., 1971). Poliovirions are not penetrated by PTA and appear as electron dense particles; in contrast, empty capsid structures, also found in infected cells and in highly purified virus preparations, are penetrated by the stain (Fig. 1).

Several investigators, using photographic reversals (in order to improve contrast) of electron micrographs of negatively stained poliovirions, have reported that the virion surface is composed of 32 capsomeres (MAYOR, 1964). However, these structures are not apparent when viewed directly even in the best negatively stained preparations. In addition, none of the picornaviruses studied to date have been degraded to yield capsomeres nor have such structures been isolated from infected cells.

Crystallographic experiments indicate that poliovirus particles possess 5:3:2 cubic symmetry (also called icosahedral symmetry; FINCH and KLUG, 1959; HORNE and WILDY, 1961). Electron microscopic examination of poliovirus *suggests* that the particle resembles most closely a rhombic triacontrahedron (MAYOR, 1964) while at least one rhinovirus seems to resemble a pentaxis dodecahedron (JAMISON, 1969). Both of these latter geometric forms exhibit icosahedral symmetry and in fact are closely related to a simple icosahedron.

B. Viral RNA

Recent measurements show that the poliovirion probably contains a single molecule of single-stranded RNA weighing approximately 2.5×10^6 daltons (GRANBOULAN and GIRARD, 1969; TANNOCK et al., 1970), a value significantly higher than the 2×10^6 daltons reported in the past (ANDERER and RESTLE, 1964). Other members of the picornavirus group, such as encephalomyocarditis (EMC) virus and certain rhinoviruses, have also been shown to possess RNA molecules of this size (BURNESS and CLOTHIER, 1970; BROWN et al., 1970). The viral genome may exist as a core within the viral capsid or it may be

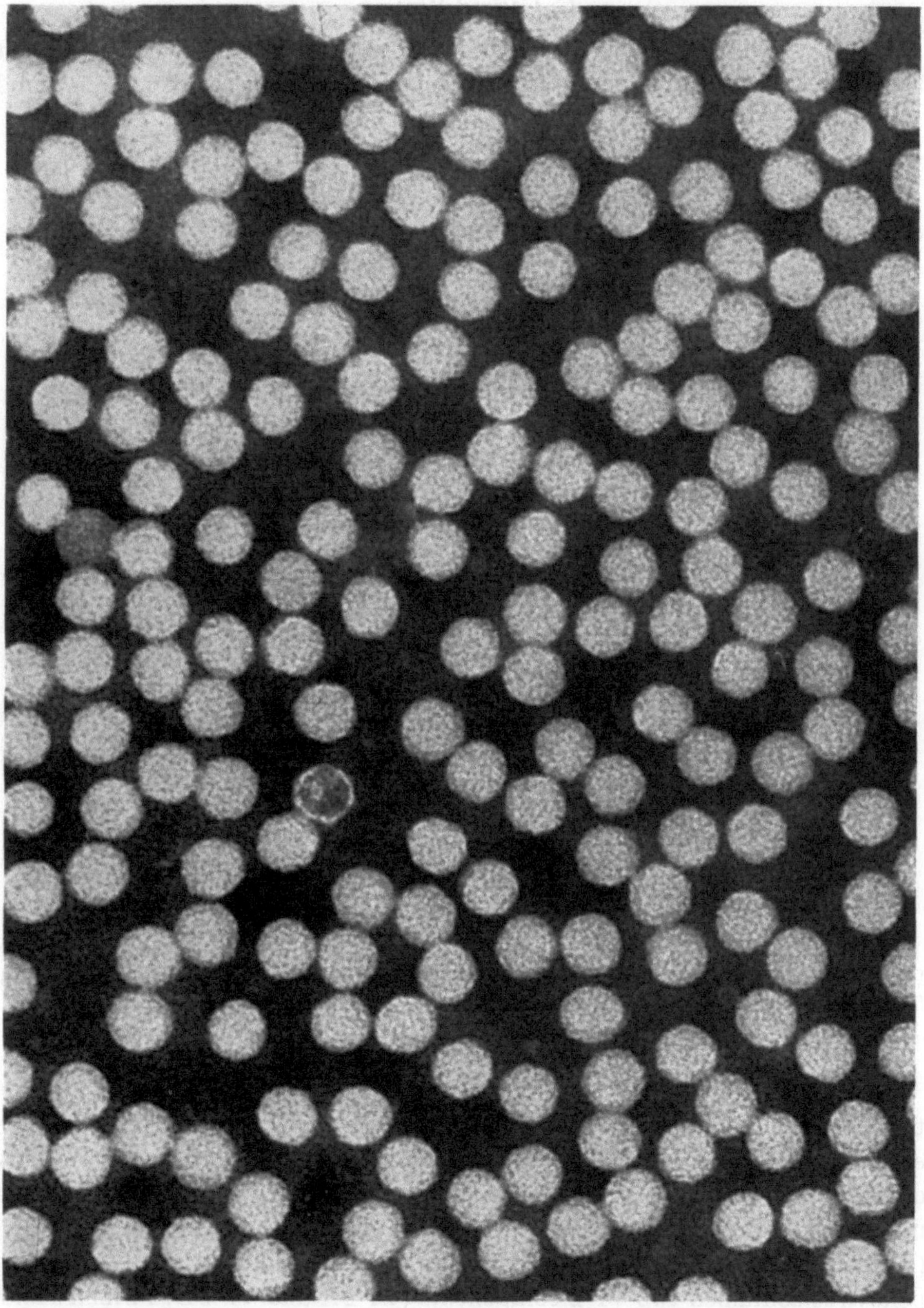

Fig. 1. Electron micrograph of highly purified type 1, Mahoney poliovirions stained with PTA, pH 7. Magnification ×300000

interpenetrated throughout the capsid. The present state of knowledge does not allow a choice between these alternatives.

C. Capsid Proteins

Highly purified poliovirus preparations, when dissociated with detergent and subjected to reduction by mercaptoethanol or dithiothreitol, were found

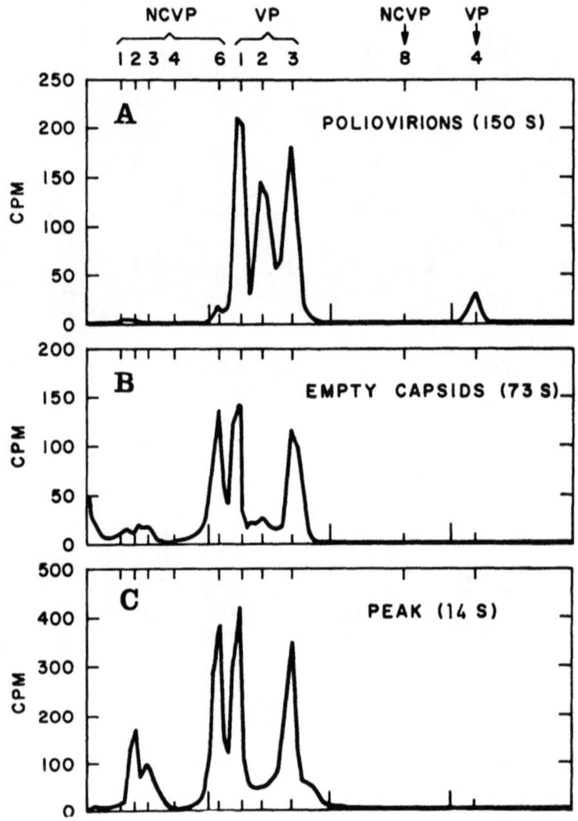

Fig. 2A–C. Acrylamide gel electropherograms of the dissociated polypeptides derived from purified poliovirions (A), partially purified naturally-occurring empty capsids (B), and partially purified 14 S particles (C). See PHILLIPS et al. (1968)

to yield at least 4 distinct polypeptide monomers which could be separated by electrophoresis in polyacrylamide gels (MAIZEL, 1963). These capsid proteins originally were named viral proteins 1 through 4 (VP 1–4; Fig. 2A) and their molecular weights were found to be 35000, 28000, 24000 and 6000, respectively (MAIZEL and SUMMERS, 1968). Later experiments showed that type 1, Mahoney poliovirus contained two VP 2 components (VP 2A and VP 2B; PHILLIPS and MAIZEL, 1967; COOPER et al., 1970). Other picornaviruses also were found to contain multiple components (RUECKERT, 1965; BURNESS and WALTER, 1967; KORANT et al., 1970). One or more of these proteins may exist as a glycoprotein (TILLOTSON and LERNER, 1966).

When poliovirions were subjected to heat and/or alkaline pH conditions to produce empty capsids, the viral RNA and the VP 4 polypeptide were lost quantitatively from the virions (MAIZEL et al., 1967)[2]. The remaining empty capsids contained VP 1, VP 2, and VP 3. Naturally-occurring empty capsids isolated from poliovirus-infected HeLa cells possess VP 1, VP 2, and VP 3 but are devoid of VP 4 (Fig. 2B). In addition, they contain a major polypeptide (called noncapsid viral protein 6 = NCVP 6) that is present in only trace amounts in the purified virion (Fig. 2B). Another difference between naturally-occurring and artificially-produced empty capsids is that the latter contain a much greater amount of the VP 2 component. These findings show that naturally-occurring empty capsids are not degradation products of the virion. Their hypothetical role in the morphogenesis of poliovirus will be discussed below.

D. Antigenicity

There are 3 distinct, non-cross-reacting, antigenic serotypes of poliovirus designated types 1, 2 and 3. In early studies involving the characterization of the virus, it was discovered that there were two non-cross-reacting antigens in crude preparations of any *given* serotype. More refined fractionation procedures were required to show that the D (or N) antigen was associated with the infectious particle while the C (or H) antigen was associated with a non-infectious particle of similar size (BACHRACH and SCHWERDT, 1954; ROIZMAN et al., 1958; BEALE, 1961; HUMMELER et al., 1962; SVEHAG and BLOTH, 1967). The C antigen, separated by its lesser density in CsCl gradients or its slower sedimentation in sucrose gradients, was found to be penetrated by negative stains and was identified as an empty capsid devoid of RNA. It has since been shown that when poliovirions are converted to empty capsids by heat (*e.g.*, 55 C, 15 min) or UV irradiation, there is an accompanying change from the D antigenicity to the C antigenicity (BOEYÉ and VAN ELSEN, 1967; KATAGIRI et al., 1967). There is some evidence that exposure of poliovirions to 35 C for 7 days results in particles which remain impermeable to negative stains yet react with anti-C serum (DIMMOCK and HARRIS, 1967).

It should be pointed out that the presence of the NCVP 6 polypeptide in naturally-occurring empty capsids does not explain its unique antigenicity because artificially-produced empty capsids, which do not possess this polypeptide, also exhibit C antigenicity. Therefore, the precise nature of the D and C antigenic determinants remains to be correlated with the various polypeptide components and it seems likely that the configuration of the polypeptides in the two kinds of particles is crucial in determining their antigenicity. In this regard, PHILLIPS and MAIZEL (1967) were able to show that the VP 1, VP 2, and VP 3 components of the three poliovirus serotypes could be distinguished by their different electrophoretic mobilities in polyacrylamide gels. The extremely rapid mobility of the VP 4 polypeptides of the different serotypes did not permit the resolution of any differences.

---

2 It is interesting that CROWELL and PHILIPSON (1971) have shown that Coxsackievirus B3 following adsorption to and spontaneous elution from susceptible cells preferentially loses the VP 4 polypeptide. The eluted particles, which contained infectious RNA molecules, no longer were able to attach to cells.

A third antigen, called the S (soluble) antigen, was isolated by SCHARFF et al. (1964) by degrading poliovirions with 6.5 M guanidine hydrochloride. This antigen has a $S_{20,w}$ of 2 S (molecular weight of 27000 daltons) and could be detected as a 5 S component in infected cells early in the course of infection.

III. Morphogenesis of the Poliovirion

A. Growth Cycle

Until the last 5 years or so, not much information was available pertinent to the maturation of poliovirus. Early studies primarily were aimed at elucidating the growth cycle of the virus in susceptible cell populations, *viz.*, determing the production of infectivity or viral-specific antigens during a single cycle of virus replication (HOWES and MELNICK, 1957; DARNELL, 1958; HOWES, 1959a, b; FURNESS, 1961; SCHARFF and LEVINTOW, 1963). Generally, the higher the multiplicity of infection (m.o.i.) the shorter the eclipse and latent periods. However, the *rate* of production of progeny virus was the same regardless of the m.o.i. (DARNELL, 1958). Progeny virus began to appear about 3 hours after attachment of the virus and maximum yields were reached by 6 to 7 hours. Although the release of virions began immediately after maturation, 95 percent or more of the progeny virus remained intracellular by 6 hrs post-infection. After 15 hours, almost all of the progeny virus was released into the medium. Since single cells were found to release progeny virus over a 1 hour span beginning at 6 to 7 hrs (LWOFF et al., 1955), infected cell populations seem to be asynchronous in this respect.

B. Poliovirus-induced Alterations in Cells

A number of investigators have employed the electron microscope as a tool for following the development of poliovirus in infected cells. Generally, the course of virus development and its effect on cells can be described as follows: At three hours post-infection, the first observed cellular change involves a separation of the nuclear membranes from one another producing a space (centrosphere) which contains nuclear and/or cytoplasmic blebs. Subsequently, the large virus-specific polyribosome clusters can be distinguished in the cytoplasm. At 4 to 5 hours, some investigators describe the appearance of a dense cytoplasmic matrix called a viroplasm. Still later, crystalline arrays of virions form in the cytoplasm possibly in those areas of the cell which contain the viroplasm. The existence of a membrane surrounding these virus crystals remains questionable. Virions are also seen in the centrosphere region as well as in cytoplasmic vacuoles. Virions are never observed in the nucleus although one study shows what might be viral empty capsids in the nucleus of an infected cell (ANZAI and OZAKI, 1969).

HORNE and NAGINGTON (1959) found a number of viral structures, including virions, empty capsids, and what appeared to be partially-filled capsids, within membraneous enclosures isolated from infected cells. Similar structures were found in ECHO virus-infected cells (SKINNER et al., 1968). In the latter case, HBB (2-α-hydroxybenzyl-benzimidazole), a specific inhibitor of viral RNA synthesis, prevented the formation of the abnormal vesicles. It is not clear from these studies how the crystalline arrays of virions are formed in the cell. Equally unclear is the nature and function of the viroplasm and its relation to the altered membrane structures seen in infected cells. It seems likely that the source of these membranes is the endoplasmic reticulum. There is no evidence to suggest a direct involvement of the mitochondria which seem to remain intact throughout the infectious cycle.

C. Time Course of Synthesis of Viral Constituents

Early studies by DARNELL and coworkers showed that viral RNA synthesis preceded viral maturation by 30 min or less (DARNELL et al., 1961). Similarly, the synthesis of viral proteins as detected by their incorporation into progeny virions was found to precede maturation by about the same time (DARNELL and LEVINTOW, 1960). Later, PENMAN et al. (1964) showed that 5–10 min were required to detect RNA in virions following a pulse of labeled nucleoside; about 20 min was required to label virions with radioactive amino acids.

SCHARFF and LEVINTOW (1963) employed the techniques of labeling virus-specific products with radioactive precursors and immune coprecipitation to follow the production of viral antigens. In these experiments, newly infected cells were exposed to either a labeled amino acid or nucleoside and, at suitable intervals, the cells were disrupted and treated with specific anti-C or anti-D serum. The immune precipitate was collected and the amount of radioactivity measured. The appearance of both D and C antigens (virions and empty capsids, respectively) was found to parallel the production of progeny virus as determined by infectivity assays. The C antigen was found to comprise from 25 to 57 percent of the total (D+C) antigenic material produced in the infected cell. Later, SCHARFF et al. (1964) were able to identify a virus-specific, soluble antigen (S antigen, see above), the time course of synthesis of this antigen suggested that it was a precursor in the formation of virions.

D. Role of Cellular Membranes in Virus Replication

The role of cellular membranes in poliovirus infection has been studied extensively by DARNELL and coworkers (BECKER et al., 1963; PENMAN et al., 1964) who isolated and characterized a cytoplasmic particulate they called a viral-synthesizing body (VSB). These large structures could be sedimented through a 10–30% w/w sucrose gradient in 90 min at 25000 rpm. Isolated VSB structures were shown to contain most of the viral-specific polyribosomes and newly synthesized protein, much if not all of the viral RNA polymerase activity, newly synthesized viral RNA molecules, and newly formed virions. These various components were released from the VSB by treatment with deoxy-cholate and subsequently re-isolated by sucrose gradient fractionation. The sedimentation behavior and other properties of the VSB suggested that they were in fact the membraneous vesicles visualized by HORNE and NAGINGTON

(1959) in their electron microscopic study of poliovirus-infected cells. PENMAN (1965) later found that one of the early effects caused by poliovirus infection was a sudden stimulation in the uptake of choline from the medium. This stimulation did not occur when cells were infected in the presence of guanidine. Presumably, the choline is incorporated into lecithin which in turn is used in the synthesis of membrane constituents. Recently, CALIGUIRI and TAMM (1969, 1970a, b) used discontinuous sucrose gradients to separate the smooth and rough endoplasmic reticulum from polio-infected HeLa cells. They found viral-specific products associated with 3 of the 6 cytoplasmic membrane fractions obtained. Recognizing that viral RNA and protein synthesis as well as viral morphogenesis are dynamic and interrelated processes, CALIGUIRI and TAMM proposed that viral RNA synthesis commences on "smooth" membranes. Upon completion of new RNA molecules, protein synthesis can begin with the binding of ribosomes to the newly made viral RNA. The resulting RNA-protein-membrane complex behaves as a "rough" membrane exhibiting different physical properties than the "smooth" membranes.

E. Role of Subviral Components

The formation of virions seems to be more intimately coupled with the synthesis of viral RNA than with the synthesis of viral proteins. For instance, radioactive uridine can be detected in virions 5 to 10 min after its addition to infected cells (see above). BALTIMORE (1968) showed that guanidine blocked the formation of virions faster than it inhibited viral RNA synthesis. Under these conditions, empty capsid structures accumulated in the infected cell (HALPEREN et al., 1964; JACOBSON and BALTIMORE, 1968a). However, the interpretation of the effect of guanidine must await an unequivocal determination of whether the drug acts directly on viral RNA synthesis or viral maturation. In contrast, once sufficient time is allowed for the synthesis of some viral protein, a viral RNA molecule can be "encapsidated" in the presence of an inhibitor of protein synthesis (BALTIMORE et al., 1966). These findings suggest that the formation of empty capsids does not require the concomitant synthesis of viral RNA whereas the formation of virions may be dependent on concurrent synthesis of the viral genome.

In the past, the empty capsids found in purified virus preparations as well as in infected cells were thought to be a degradation product of the poliovirion. Other studies suggested that the empty capsid was an end-product resulting from the assembly of protein monomers into a capsid in the absence of viral nucleic acid (SCHARFF and LEVINTOW, 1963; HALPEREN et al., 1964a, b). Later, MAIZEL et al. (1967) showed that naturally-occurring empty capsids possessed a major polypeptide not present (or present in only trace amounts) in the virion. This finding ruled out the possibility that empty capsids were derived from virions. JACOBSON and BALTIMORE (1968a) showed that if guanidine were added to infected cells midway through the infectious cycle, viral RNA synthesis abruptly stopped but empty capsids began to accumulate in the cell. When the guanidine subsequently was removed by washing, the intracellular concentration of the empty capsids, already made radioactive by a pulse of labeled amino acids, declined and radioactive virions appeared within 5 to 10 min. The appearance of radioactivity in the virions under these conditions was sooner than the 20 to 30 min normally required following a pulse of radioactive amino acids (see above). Since the removal of guanidine was conducted in a medium containing a high concentration of unlabeled amino acids (to prevent the further incorporation of labeled amino acids into viral specific products), the results were interpreted as showing that empty capsids had become virions upon the resumption of viral RNA synthesis. Unfortunately, these experiments did not rule out the possibility that the empty capsids were degraded to some intermediate structure(s) which were then reassembled into virions. In spite of this, these findings seem most compatible with the idea that empty capsids are precursors of the virion.

In addition to the 150S virion and 73S empty capsid, WATANABE et al. (1962) first described a labeled 10S component in infected cells that had been exposed to radioactive amino acids. The 10S material reacted with antiserum prepared against intact virions (WATANABE et al., 1965). However, this antiserum also reacted with empty capsids. On the other hand, antiserum specific for the 10S material did not cross-react with virions (D antigen) or empty capsids (C antigen). Thus, it appears that the 10S component possesses a unique antigenic specificity. Its relationship to the S antigen described by SCHARFF and coworkers (see abowe) remains to be elucidated.

Independently, PHILLIPS et al. (1967, 1968) also discovered a 10S particle, the sedimentation of which was later corrected to 13–14S (molecular weight estimated to be approximately 400000 daltons). Only 5–10 min were required to label the 14S particles following the addition of radioactive amino acids to infected cells. By analyzing the dissociated polypeptides on polyacrylamide gels, the 14S particles were found to contain the same polypeptides as naturally-occurring empty capsids (see Fig. 2C). Furthermore, when labeled 14S particles were added to extracts prepared from infected HeLa cells and incubated at 37C, about 40% of the recovered radioactivity sedimented as 73S material (PHILLIPS et al., 1968). No such conversion resulted when extracts prepared from uninfected cells were used. The sedimentation properties, sensitivity to SDS, and polypeptide content of the 73S particles indicated that they were similar, if not identical, to naturally-occurring empty capsids. These findings suggested that the 14S particles were precursors in the formation of empty capsids. It was also found that: (i) the assembly activity of infected cell extracts was sensitive to trypsin and deoxycholate but not to metabolic poisons such as CN^- or F^- or protein inhibitors such as puromycin and cycloheximide; (ii) neither viral RNA molecules nor viral RNA syntheses *per se* were necessary for the assembly activity; (iii) the assembly activity was maximal at mid-cycle infection, and (iv) the endogenous supply of 14S particles was not responsible for the assembly activity (PHILLIPS, 1969).

Later, it was discovered that 14S particles were able to self-assemble into a 73S particle which upon electron microscopic examination was in fact an empty capsid (PHILLIPS, 1970, 1971). The self-assembly reaction, in marked contrast to the extract-mediated reaction, was found to be extremely concentration dependent; a 2-fold dilution in buffer resulted in a 50% inhibition of 73 S particle formation. In the absence of any dilution, the initial kinetics of 14 S to 73 S conversion was identical to the kinetics measured in the presence of infected cell extracts; in spite of the fact that in the latter experiments the 14 S particles were diluted by about 10-fold.

It is clear from these results that infected cell extracts contain some factor(s) which permit the assembly reaction to occur under conditions (of dilution) that would otherwise result in little or no assembly of 14 S particles into empty capsids. The nature of this activity remains unknown although its susceptibility to destruction by trypsin and deoxycholate (PHILLIPS, 1969) and its sedimentation behavior (unpublished data) suggest that it is associated with membranes.

GHENDON and coworkers (1970) studied the formation of viral products in cells infected with a temperature-sensitive (ts) mutant of type 2 poliovirus. In this system, infection at 40C (nonpermissive conditions) was accompanied by inhibition of host cell macromolecular synthesis but there was no detectable replication of viral RNA and no measurable viral RNA replicase activity. Infected cells incubated at 40C did synthesize 14 S particles although at markedly reduced levels. However, neither virions nor empty capsids were formed. In addition to the 14 S particles, material sedimenting at 5 S and 2 S (also see PHILLIPS et al., 1968) was produced under the nonpermissive conditions. A further study by MIKHEJEVA et al. (1970) made use of a number of temperature-sensitive mutants of type 1, Mahoney poliovirus to study viral morphogenesis. These mutants were grouped into 3 classes based on their ability to synthesize viral RNA and viral antigen under nonpermissive conditions. One class failed to synthesize either viral RNA or viral antigen. A second class failed to synthesize detectable viral RNA but did synthesize viral antigen. The third class of mutants produced both viral RNA (38–58% of wild-type) and viral antigen. Under nonpermissive conditions, none of the mutants produced virions or empty capsids. Mutants unable to synthesize detectable viral RNA produced only the 5 S component, while the mutants able to synthesize RNA produced both the 14 S and 5 S components. However, only one mutant of each class was tested for the nature of its morphogenetic defect.

These results seem to support the idea that both 5 S and 14 S components are precursors in the formation of poliovirions. In addition, the fact that certain mutants direct the synthesis of 14 S particles but not empty capsids may be explained by: (i) the formation of defective 14 S particles; (ii) an insufficient supply of 14 S particles; or (iii) a defect in the assembly activity demonstrable in infected, cytoplasmic extracts. Unfortunately, the correct answer must await further experimentation with these mutants. The finding that 5 S and 14 S material was synthesized in cells infected with ts mutants under nonpermissive conditions suggests that these products are made from the parental

RNA. However, as pointed out by the authors, the detection of viral RNA synthesis at levels below 10% of that of wild-type virus is not possible. Therefore, in mutants designated as RNA⁻, limited viral RNA synthesis may indeed occur thus accounting for the formation of certain viral proteins and detectable levels of certain subviral components.

F. Cleavage of Viral Proteins

It was mentioned earlier that poliovirions are composed of at least 5 structural polypeptides. SUMMERS et al. (1965) were the first to show that at least 10 additional virus-specific polypeptides are present in infected cells (Fig. 3); similar results have been obtained for other members of the picornavirus family (HOLLAND and KIEHN, 1968). These findings created a dilemma in that the number of virus-specific polypeptides exceeded by a factor of 2 the genetic coding capacity of the viral RNA. Later, it was found that not all of the polypeptides detected in infected cells represented primary gene products (SUMMERS and MAIZEL, 1968; JACOBSON and BALTIMORE, 1968b). Evidence was obtained that certain of the viral polypeptides (*e.g.* VP 2 and VP 3) were produced by the cleavage of larger polypeptides. This was first demonstrated by electrophoresis in polyacrylamide gels of the labeled viral polypeptides produced in infected cells at various times following a 3 min pulse with radioactive amino acids and chase with excess unlabeled amino acids (see Fig. 4). A short pulse of radioactive amino acids labeled NCVP 1 whereas the structural polypeptides VP 1, VP 2 and VP 3 showed little or no radioactivity. However, during the chase, the radioactivity of NCVP 1 diminished greatly while the structural polypeptides became more radioactive. The NCVP 2 component was rapidly labeled during the pulse but its radioactive content was stable during the chase period. These findings were quickly verified by JACOBSON and BALTIMORE (1968b). In addition, these investigators found that the NCVP 6 polypeptide[3] was cleaved to form VP 2 when the empty capsid and RNA combined to form the virion (JACOBSON and BALTIMORE, 1968a). It should be noted that there is some disagreement in whether VP 2 exists in the empty capsid. JACOBSON and coworkers did not find significant amounts in empty capsids whereas MAIZEL et al. (1967), and PHILLIPS et al. (1968), detected small but significant amounts of VP 2 in empty capsids. In addition, VP 2 seems to be present in low but significant amounts in the 14S particles which are precursors to the empty capsids (PHILLIPS, 1971). The fact that there are two VP 2 polypeptides in some or all of the poliovirus strains may be important; it is possible that one of them is present in 14S and 73S empty capsids and the other is produced by the cleavage of NCVP 6 when empty capsids combine with the viral RNA.

Later, JACOBSON et al. (1970) showed that the incorporation of amino acid analogs prevented or slowed down the cleavage reaction(s). By adding combinations of analogs to infected cells these investigators were able to cause the accumulation of large molecular weight polypeptides, some of which

3 Called VP 0 by these workers.

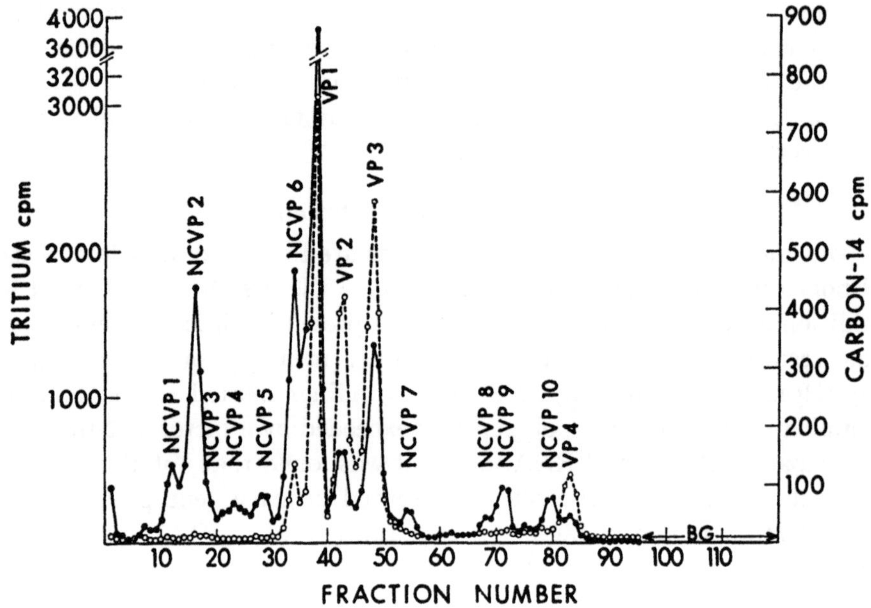

Fig. 3. Poliovirus-specific proteins in total cytoplasmic extract of infected cells. Twenty-five ml of infected cells were guanidine-treated for 2 hr, the guanidine washed out, and the cells resuspended in medium containing H^3 amino acids. The cytoplasmic extract was prepared at 7 hrs post-infection and mixed with purified virus labeled with C^{14} amino acids. After dissociation and reduction, the mixture was electrophoresed in acrylamide gels. C^{14}—●—, H^3--○--. (From SUMMERS et al., 1965; data reproduced with permission of the National Academy of Sciences)

were not detectable under normal conditions of labeling *in vivo*. One of these components, called NCVP 00, was calculated to have a molecular weight of greater than 200000 daltons; such a protein might represent a single translation product of the entire viral genome. Presumably, the incorporation of analogs changes the configuration of the polypeptide chain so that the sites of cleavage are altered or made inaccessible to the cleaving enzyme(s). Assuming the cleavage reactions are enzymic, the fact that they occur in different cells infected with a number of different picornaviruses (HOLLAND and KIEHN, 1968) suggests that either these enzymes are encoded in the genome of the infecting virus or they are present ubiquitously in mammalian cells.

COOPER et al. (1970) were able to detect electrophoretic differences in polyacrylamide gels between some of the structural polypeptides of closely related strains and mutants of type 1 poliovirus. For example, the VP 1 polypeptides of type 1, Mahoney and type 1, LSc (Sabin vaccine strain derived from type 1, Mahoney) virions differed by about 5000 daltons yet produced remarkably similar elution patterns when their respective tryptic peptides were analyzed by ion-exchange chromatography. To account for these findings, COOPER et al. proposed that mutations in the viral RNA might affect the number and/or relative position of cleavage sites by either of two mechanisms:

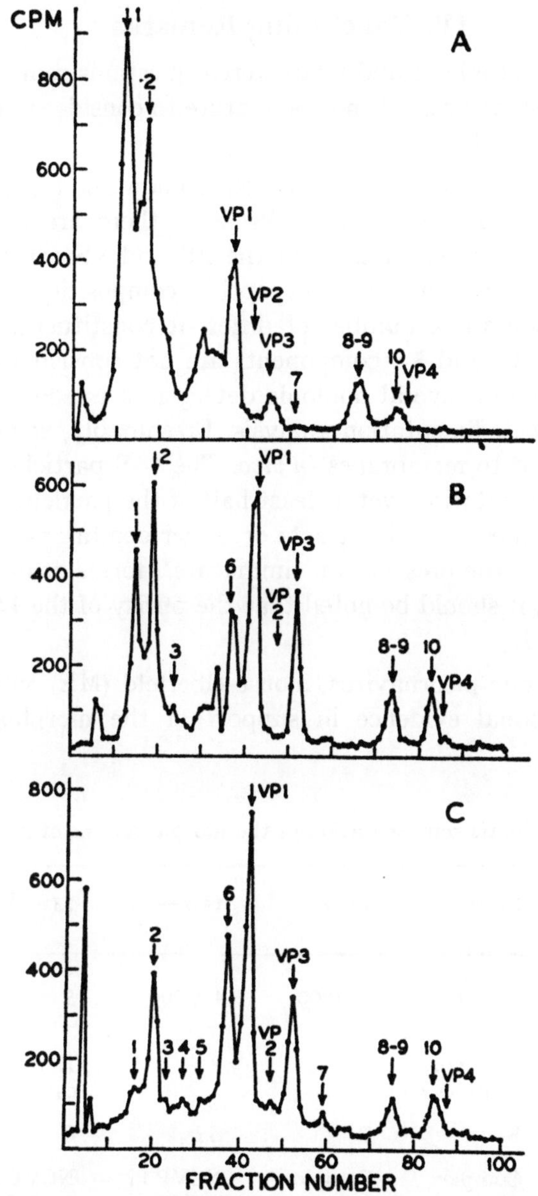

Fig. 4 A–C. Progressive changes in the pattern of radioactivity in the cytoplasmic proteins of poliovirus-infected HeLa cells (3 min pulse with C^{14} amino acids) followed by a chase with C^{12} amino acids for 2 min (A), 10 min (B), and 45 min (C). Cytoplasmic extracts were prepared from fractions of the cell culture at each of the chase times, subjected to dissociation and reduction, and analyzed by acrylamide gel electrophoresis. Noncapsid polypeptides are designated as 1 through 10, and virus capsid polypeptides as VP 1 through VP 4. (From MAIZEL and SUMMERS, 1968; data reproduced with permission of the National Academy of Sciences)

(i) directly because of a substituted amino acid at the cleavage site, or (ii) indirectly as a result of a configurational change in the entire precursor polypeptide.

IV. Concluding Remarks

Although many problems and unanswered questions remain regarding the morphogenesis of poliovirus, it is not premature to consider a model which can serve as a useful basis for further experiments.

Table 1 contains "everything you wanted to know about poliovirus morphogenesis, but were afraid to ask". To be sure, there are many speculative elements in this scheme. The nature of the 2S and 5S components remains largely unknown. Their complex polypeptide composition almost certainly means that they contain a number of different constituents co-sedimenting together. Also, the 2S and 5S components are not consistently found unless deoxycholate (or one of several nonionic detergents) is added to the infected cell extract prior to sedimentation analysis. Presumably, some of these components are attached to membranes *in vivo*. The 14S particle is homogeneous with regard to sedimentation, yet at least half of the particles fail to assemble in the presence of infected cell extracts or under conditions of self-assembly. Whether this reflects the presence of "immature" forms or damaged particles is not known. Also, it should be noted that the purity of the 14S particles has not been established.

In the case of one picornavirus, mouse-elberfeld (ME) virus, degradation studies lend additional evidence in support of the morphogenetic process

Table 1. Schematic representation of the morphogenesis of poliovirions

Particles size ($S_{20,w}$)	2S →	5S →	14S →	$73S \xrightarrow{+35S\ RNA}$	150S
Molecular weight (daltons)	27000	100000	400000	5.0×10^6	7.5×10^6
Morphology	Free polypeptides?	?	?	Empty capsid	Virion
Antigenicity	S	S	S?	C	D
Polypeptide composition	Complex	Complex[a]	NCVP 1? NCVP 6 VP 1 VP 2 VP 3	NCVP 6 VP 1 VP 2 VP 3	VP 1 VP 2 VP 3 VP 4[b]
Cleavage reactions	NCVP 1 → NCVP 6, VP1, VP 3			NCVP 6→VP 2+(VP4?)	

[a] Electrophoretic pattern resembles that obtained from infected, cytoplasmic extracts (see PHILLIPS et al., 1968)

[b] Recently identified as being the D-reactive part of the poliovirion [BREINDL; Virology 46, 962–964 (1971)].

described in Table 1. RUECKERT et al. (1969) showed that treatment of ME virus with mild acid produced a 14 S particle. This particle, in turn, could be degraded to a 5 S particle by brief exposure to 2M urea. From these data and the results of PHILLIPS et al. (1968) they proposed that the 14 S particle was composed of five 5 S "monomers" and that twelve 14 S particles aggregated to form an empty capsid.

It is clear that the 14 S particle is too large to be a capsomere; the molecular weight of this particle is more consistent with a dimeric capsomere if such a structure indeed exists. And the 5 S particle is too small to be a capsomere; the molecular weight of 32 5S particles would be only 3.2×10^6 (see Table 1). On the other hand, the aggregation of twelve 14S particles, assuming no gain or loss of polypeptide constituents, would result in a particle weighing about 4.8×10^6 daltons. This value is in reasonable accord with the theoretical particle weight of the empty capsid.

The significance of the assembly of 14 S particles into empty capsids relies, in part, on the role of the latter in the morphogenesis of the poliovirion. The experimental results to date rule out the possibility that naturally-occurring empty capsids are degradation products of the virion and favor, but do not prove unequivocally, their role as a direct precursor of the virion. The other possible explanation is that empty capsids are *end-products* resulting from the aggregation of 14S particles in the absence of the viral genome. Admittedly, it is difficult to envision the mechanism whereby an intact viral RNA molecule combines with an empty capsid. However, two important considerations bear on this problem. First, a major alteration in the surface structure accompanies this combination as indicated by (i) a cleavage of a major polypeptide (NCVP 6 → VP 2), and (ii) a change in the antigenic nature (C → D) of the particle. Second, based on studies of turnip yellow mosaic virus (TYMV), a plant virus whose morphology is identical to poliovirus, the viral RNA does not appear to be a tightly coiled ball in the center of the virion. Rather, the RNA seems to be interspersed throughout the capsid protein (KLUG and FINCH, 1960; also see CASPAR and KLUG, 1962). In fact, portions of the viral RNA may be exposed at the virion surface but complexed in such a way as to be protected from nucleases. Thus, the idea of "stuffing" an RNA molecule into an empty capsid is neither appealing nor necessary. Given the right environmental conditions, the RNA may integrate with the capsid surface forming a stable structure.

Modified endoplasmic reticulum appears to play an important role in the morphogenesis of poliovirus. These altered membranes may provide a physical site at which viral RNA and proteins are synthesized. Considering the size of the cell and the number and size of the viral constituents, it is logical to suggest that the retention of newly synthesized viral components at a given locus is necessary for their later interaction. This is consistent with the repeated finding that the machinery for synthesizing viral RNA and proteins is associated with membranes. The fact that the self-assembly of 14S particles into empty capsids is a concentration-dependent reaction whereas no such concentration dependence is evident in the presence of infected cell extracts suggests that the extracts "facilitate" the self-assembly reaction rather than supply a needed factor. To explain these findings, it has been proposed that a modified membrane surface is needed to bind 14S particles to specific sites thus creating the conditions favorable for their self-assembly (PHILLIPS, 1971). In a sense, the membrane assembly sites would act in a manner more subtle than usually envisioned for a maturation factor.

In conclusion, the poliovirion is a small, spherical ribonucleoprotein particle whose relatively simple morphology belies a surprisingly complex morphogenetic mechanism. Presumably, the self-assembly of viral subunits plays a role in the formation of certain subviral particles. The process is complicated by the fact that viral subunits are composed of multiple polypeptides some of which undergo cleavage reactions during the assembly process. Perhaps the most intriguing question is how the viral RNA combines with an empty capsid, presuming a precursor role for the latter particle; the cleavage of the NCVP 6 polypeptide at the moment of this combination may be crucial.

References

ANDERER, V. F. A., RESTLE, H.: Untersuchungen über ein attenuiertes Poliomyelitis-Virus Typ II, Reindarstellung und physikalisch-chemische Eigenschaften des Virus. Z. Naturforsch. **19**b, 1026–1031 (1964).

ANZAI, T., OZAKI, Y.: Intranuclear crystal formation of poliovirus: Electron microscopic observations. Exp. molec. Path. **10**, 176–185 (1969).

BACHRACH, H. L., SCHWERDT, C. E.: Purification studies on Lansing poliomyelitis virus. J. Immunol. **72**, 30—38 (1954).

BALTIMORE, D.: Inhibition of poliovirus replication by guanidine. *In:* Medical and applied virology. Proc. of the 2nd Internatl. Symp. (eds. SANDERS, M., and LENNETTE, E. M.), p. 340. St. Louis: Green 1968.

BALTIMORE, D., GIRARD, M., DARNELL, J. E.: Aspects of the synthesis of poliovirus RNA and the formation of virus particles. Virology **29**, 179–189 (1966).

BEALE, A. J.: The D-antigen content in poliovaccine as a measure of potency. Lancet **1961** II, 1166–1168.

BECKER, Y., PENMAN, S., DARNELL, J. E.: A cytoplasmic particulate involved in poliovirus synthesis. Virology **21**, 274–276 (1963).

BOEYÉ, A., VAN ELSEN, A.: Alkaline disruption of poliovirus: Kinetics and purification of RNA-free particles. Virology **33**, 335–343 (1967).

BROWN, F., NEWMAN, J. F. E., STOTT, E. J.: Molecular weight of rhinovirus ribonucleic acid. J. gen. Virol. **8**, 145–148 (1970).

BURNESS, A. T. H., CLOTHIER, F. W.: Particle weight and other biophysical properties of encephalomyocarditis virus. J. gen. Virol. **6**, 381–393 (1970).

BURNESS, A. T. H., WALTER, D. S.: Protein components of encephalomyocarditis virus. Nature (Lond.) **215**, 1350–1352 (1967).

CALIGUIRI, L., TAMM, I.: Membraneous structures associated with translation and transcription of poliovirus RNA. Science **166**, 885–886 (1969).

CALIGUIRI, L., TAMM, I.: The role of cytoplasmic membranes in poliovirus biosynthesis. Virology **42**, 100–111 (1970a).

CALIGUIRI, L., TAMM, I.: Characterization of poliovirus-specific structures associated with cytoplasmic membranes. Virology **42**, 112–122 (1970b).

CASPAR, D. L. D., KLUG, A.: Physical principles in the construction of regular viruses. Cold Spr. Harb. Symp. quant. Biol. **27**, 1–24 (1962).

COOPER, P. D., SUMMERS, D. F., MAIZEL, J. V.: Evidence for ambiguity in the post-translational cleavage of poliovirus proteins. Virology **41**, 408–418 (1970).

CROWELL, R. L., PHILIPSON, L.: Properties of Coxsackievirus B3 eluted from HeLa cells. Bact. Proc. p. 177 (1971).

DALES, S., EGGERS, H. J., TAMM, I., PALADE, G. E.: Electron microscopic study of the formation of poliovirus. Virology **26**, 379–389 (1965).

DARNELL, J. E., JR.: Adsorption and maturation of poliovirus in singly and multiply infected HeLa cells. J. exp. Med. **107**, 633–641 (1958).

DARNELL, J. E., JR., LEVINTOW, L.: Poliovirus protein: Source of amino acids and time course of synthesis. J. biol. Chem. **235**, 74–77 (1960).

DARNELL, J. E., JR., LEVINTOW, L., THOREN, M. M., HOOPER, J. L.: The time course of synthesis of poliovirus RNA. Virology **13**, 271–279 (1961).

DIMMOCK, N. J., HARRIS, W. J.: Heterogeneity of apparently empty poliovirus particles with phosphotungstic acid after heating of the virus at "low" temperature. Virology **31**, 715–719 (1967).

FINCH, J. T., KLUG, A.: Structure of poliomyelitis virus. Nature (Lond.) **183**, 1709–1714 (1959).

FURNESS, G.: The effect of environment on the replication of poliovirus in monkey kidney cells. J. gen. Microbiol. **25**, 421–428 (1961).

GHENDON, Y. Z., BABUSHKINA, L., MIKHEEVA, A., SOLOVIEV, G. Y. A.: Synthesis of the virus-specific proteins of a conditionally lethal mutant of poliovirus under non-permissive conditions. Virology **40**, 595–604 (1970).

GRANBOULAN, N., GIRARD, M.: Molecular weight of poliovirus ribonucleic acid. J. Virol. **4**, 475–479 (1969).

HALPEREN, S., EGGERS, H. J., TAMM, I.: Complete and coreless hemagglutinating particles produced in ECHO 12 virus-infected cells. Virology **23**, 81–89 (1964a).

HALPEREN, S., EGGERS, H. J., TAMM, I.: Evidence for uncoupled synthesis of viral RNA and viral capsids. Virology **24**, 36–46 (1964b).

HÖGLUND, S.: Electron microscopic studies on some virus and immunoglobulin components. Ark. Kemi **28**, 505–538 (1968).

HOLLAND, J. J., KIEHN, E. D.: Specific cleavage of viral proteins as steps in the synthesis and maturation of enteroviruses. Proc. nat. Acad. Sci. (Wash.) **60**, 1015–1022 (1968).

HORNE, R. W., NAGINGTON, J.: Electron microscope studies of the development and structure of poliomyelitis virus. J. molec. Biol. **1**, 333–338 (1959).

HORNE, R. W., WILDY, P.: Symmetry in virus architecture. Virology **15**, 348–373 (1961).

HOWES, D. W.: The growth cycle of poliovirus in cultured cells. II. Mutation and release of virus in suspended cell populations. Virology **9**, 96–109 (1959a).

HOWES, D. W.: The growth cycle of poliovirus in cultured cells. III. The asynchronous response of HeLa cells multiply infected with type 1 poliovirus. Virology **9**, 110–126 (1959b).

HOWES, D. W., MELNICK, J. L.: The growth cycle of poliovirus in monkey kidney cells. I. Maturation and release of virus in monolayer cultures. Virology **4**, 97–108 (1957).

HUMMELER, K., ANDERSON, T. F., BROWN, R. A.: Identification of poliovirus particles of different antigenicity by specific agglutination as seen in the electron microscope. Virology **16**, 84–90 (1962).

JACOBSON, M. F., ASSO, J., BALTIMORE, D.: Further evidence on the formation of poliovirus proteins. J. molec. Biol. **49**, 657–669 (1970).

JACOBSON, M. F., BALTIMORE, D.: Morphogenesis of poliovirus. I. Association of the viral RNA with coat protein. J. molec. Biol. **33**, 369–378 (1968a).

JACOBSON, M. F., BALTIMORE, D.: Polypeptide cleavages in the formation of poliovirus proteins. Proc. nat. Acad. Sci. (Wash.) **61**, 77–84 (1968b).

JAMISON, R. M.: Morphology of echovirus 22. J. Virol. **4**, 904–906 (1969).

KATAGIRI, S., HIMUMA, Y., ISHIDA, N.: Biophysical properties of poliovirus particles irradiated with ultraviolet light. Virology **32**, 337–343 (1967).

KLUG, A., FINCH, J. T.: The symmetrus of the protein and nucleic acid in turnip yellow mosaic virus: X-ray diffraction studies. J. molec. Biol. **2**, 201–215 (1960).

KORANT, B. D., LONBERG-HOLM, K. K., HALPEREN, S.: Structural polypeptides of three rhinoviruses. Biochem. biophys. Res. Commun. **41**, 477–481 (1970).

LEVINTHAL, J. D., DUNNEBACKE, T. H., WILLIAMS, R. C.: Study of poliovirus infection of human and monkey cells by indirect immunoferritin technique. Virology **39**, 211–223 (1969).

LWOFF, A., DULBECCO, R., VOGT, M., LWOFF, M.: Kinetics of the release of poliomyelitis virus from single cells. Virology **1**, 128–139 (1955).

Maizel, J. V., Jr.: Evidence for multiple components in the structural protein of type 1 poliovirus. Biochem. biophys. Res. Commun. **13**, 483–489 (1963).

Maizel, J. V., Jr., Phillips, B. A., Summers, D. F.: Composition of artificially produced and naturally occurring empty capsids of poliovirus type 1. Virology **32**, 692–699 (1967).

Maizel, J. V., Jr., Summers, D. F.: Evidence for differences in size and composition of the poliovirus-specific polypeptides in infected HeLa cells. Virology **36**, 48–54 (1968).

Mattern, C. F. T., Daniel, W. A.: Replication of poliovirus in HeLa cells: Electron microscopic observations. Virology **26**, 646–663 (1965).

Mayor, H. D.: Picornavirus symmetry. Virology **22**, 156–160 (1964).

McFerran, J. B., Clarke, J. K., Connor, T. J.: The size of some mammalian picornaviruses. J. gen. Virol. **10**, 279–284 (1971).

Mikhejeva, A., Yakobson, E., Soloviev, G. Y.: Characterization of some poliovirus temperature-sensitive mutants and poliovirus-related particle formation under nonpermissive conditions. J. Virol. **6**, 188–193 (1970).

Penman, S., Becker, Y., Darnell, J. E.: A cytoplasmic structure involved in the synthesis and assembly of poliovirus components. J. molec. Biol. **8**, 541–555 (1964).

Phillips, B. A.: *In vitro* assembly of polioviruses. I. Kinetics of the assembly of empty capsids and the role of extracts from infected cells. Virology **39**, 811–821 (1969).

Phillips, B. A.: *In vitro* assembly of poliovirus. II. Evidence for the self-assembly of 14 S particles into empty capsids. Virology **44**, 307–316 (1971).

Phillips, B. A., Maizel, J. V., Jr.: A comparison of the structural peptides of poliovirus types 1, 2, and 3. Fed. Proc. **26**, 482 (1967).

Phillips, B. A., Summers, D. F., Maizel, J. V., Jr.: Assembly of poliovirus related particles. Bact. Proc. p. 141 (1967).

Phillips, B. A., Summers, D. F., Maizel, J. V., Jr.: *In vitro* assembly of poliovirus-related particles. Virology **35**, 216–226 (1968).

Roizman, B., Mayer, M. M., Rapp, H. J.: Immunochemical studied of poliovirus. III. Further studies on the immunologic and physical properties of poliovirus particles produced in tissue culture. J. Immunol. **81**, 419–425 (1958).

Rueckert, R. R.: Studies on the structure of viruses of the columbia SK group. II. The protein subunits of ME-virus and other members of the columbia SK group. Virology **26**, 345–358 (1965).

Rueckert, R. R., Dunker, A. K., Stoltzfus, C. M.: The structure of mouse-elberfeld virus: A model. Proc. nat. Acad. Sci. (Wash.) **62**, 912–919 (1969).

Scharff, M. D., Levintow, L.: Quantitative study of the formation of poliovirus antigens in infected HeLa cells. Virology **19**, 491–500 (1963).

Scharff, M. D., Maizel, J. V., Jr., Levintow, L.: Physical and immunological properties of a soluble precursor of the poliovirus capsid. Proc. nat. Acad. Sci. (Wash.) **51**, 329–337 (1964).

Skinner, M. S., Halperen, S., Harkin, J. C.: Cytoplasmic membrane-bound vesicles in Echovirus 12 infected cells. Virology **36**, 241–253 (1968).

Summers, D. F., Maizel, J. V., Jr.: Evidence for large precursor proteins in poliovirus synthesis. Proc. nat. Acad. Sci. (Wash.) **59**, 966–971 (1968).

Summers, D. F., Maizel, J. V., Jr., Darnell, J. E., Jr.: Evidence for virus-specific noncapsid proteins in poliovirus-infected HeLa cells. Proc. nat. Acad. Sci. (Wash.) **54**, 505–513 (1965).

Svehag, S. E., Bloth, B.: Examination of virus-M antibody complexes by electron microscopy. I. Configuration of M antibodies and mode of attachment to antigen. Virology **31**, 676–687 (1967).

Tannock, G. A., Gibbs, A. J., Cooper, P. D.: A re-examination of the molecular weight of poliovirus RNA. Biochem. biophys. Res. Commun. **38**, 298–304 (1970).

Tillotson, J. R., Lerner, A. M.: Effect of periodate oxidation on hemagglutinating and antibody-producing capacities of certain enteroviruses and reoviruses. Proc. nat. Acad. Sci. (Wash.) **56**, 1143–1150 (1966).

Watanabe, Y., Watanabe, K., Hinuma, Y.: Synthesis of poliovirus-specific proteins in HeLa cells. Biochim. biophys. Acta (Amst.) **61**, 976–977 (1962).

Watanabe, Y., Watanabe, K., Katagiri, S., Hinuma, Y.: Virus-specific proteins produced in HeLa cells infected with poliovirus: Characterization of subunit-like protein. J. Biochem. **57**, 733–741 (1965).

Author Index

Page numbers in *italics* refer to the references

Subject Index

Index to Volumes 40—57

Volume 46

Volume 47

Volume 48

Volume 49

Volume 50

Volume 51

Volume 52

Volume 53

Volume 54

Volume 55

Volume 56

Volume 57

Current Topics in Microbiology and Immunology

Reprint from Vol. 58

Applications of Bound Biopolymers in Enzymology and Immunology

R. C. Boguslaski, R. S. Smith, N. S. Mhatre

Springer-Verlag Berlin · Heidelberg · New York 1972

Reprint from

Current Topics in Microbiology and Immunology

Vol. 58

Virus Sensitization

M. Majer

Springer-Verlag Berlin · Heidelberg · New York 1972

Reprint from

Current Topics in Microbiology and Immunology

Vol. 58

Antigenic Determinants of Proteins of Defi
Sequences

E. Benjamini, D. Michaeli, J. D. Young

Springer-Verlag Berlin · Heidelberg · New York 1972

Current Topics in Microbiology and Immunology

Reprint from Vol. 58

Replication and Lysogeny with Phage P22 i

Salmonella Typhimurium

M. Levine

Springer-Verlag Berlin · Heidelberg · New York 1972

Current Topics in Microbiology and Immunology

Reprint from Vol. 58

The Morphogenesis of Poliovirus

B. A. Phillips

Springer-Verlag Berlin · Heidelberg · New York 1972

Universitätsdruckerei H. Stürtz AG, Würzburg